Philosophical Foundations of Neurolaw

Philosophical Foundations of Neurolaw

Martin Roth

LEXINGTON BOOKS
Lanham • Boulder • New York • London

Published by Lexington Books
An imprint of The Rowman & Littlefield Publishing Group, Inc.
4501 Forbes Boulevard, Suite 200, Lanham, Maryland 20706
www.rowman.com

Unit A, Whitacre Mews, 26-34 Stannary Street, London SE11 4AB

British Library Cataloguing in Publication Information Available

Library of Congress Cataloging-in-Publication Data

Names: Roth, Martin (Associate Professor of Philosophy), author.
Title: Philosophical foundations of neurolaw / Martin Roth.
Description: Lanham : Lexington Books, 2017. | Includes bibliographical references and index.
Identifiers: LCCN 2017046425 (print) | LCCN 2017044081 (ebook) | ISBN 9781498539678 (Electronic) | ISBN 9781498539661 (cloth : alk. paper)
Subjects: LCSH: Medical laws and legislation. | Neurosciences–Philosophy. | Law–Psychological aspects. | Law–Philosophy.
Classification: LCC K3601 (print) | LCC K3601 .R67 2017 (ebook) | DDC 344.04/1–dc23
LC record available at https://lccn.loc.gov/2017046425

∞™ The paper used in this publication meets the minimum requirements of American National Standard for Information Sciences—Permanence of Paper for Printed Library Materials, ANSI/NISO Z39.48-1992.

Printed in the United States of America

For Porter

Contents

Acknowledgments

This book grew out of a course on neuroscience and law that I first taught in 2010. I thank Drake University for encouraging me to develop this course, as do I the many students who have made the class so rewarding to teach. Special thanks go to Jeremy Karlin for the valuable legal insights he provided at the outset of this project, and to Drake's Humanities Center for the support it gave to complete it. Finally, I'd like to thank Rob Cummins for being a terrific mentor and research partner.

Chapters 5 and 6 contain material drawn from previously written work, including "Two Tales of Functional Explanation" (with Robert Cummins. *Philosophical Psychology* 27 (6), 2014. Reprinted by permission of Taylor & Francis Ltd.), "Folk Psychology as Science" (*Synthese* 190 (17), 2012. With permission of Springer), "Representation, Philosophical Issues about" (*Wiley Interdisciplinary Reviews: Cognitive Science* 1 (1), 2009. Reprinted by permission of John Wiley and Sons), and "Neuroscience, Psychology, Reduction, and Functional Analysis" (with Robert Cummins. Forthcoming in *Explanation and Integration in Mind and Brain Science*, edited by David Kaplan. With permission of Oxford University Press)

Chapter 1

Two Images

The last two decades have seen an explosion of interest in what impact, if any, our growing knowledge of the brain will or should have for legal theory and practice. President George H. W. Bush designated the 1990s "the decade of the brain," and a number of initiatives, including the MacArthur Foundation's research network on law and neuroscience,[1] have been undertaken in this area. Newer technologies will promise to reveal even more about the biological basis of human behavior, and this increase in knowledge will undoubtedly prompt further consideration of the implications of such knowledge for law.[2]

The field of study that has developed around this interest has been often dubbed "neurolaw," and what makes neurolaw especially inviting to philosophers is that it is a concrete instance of the challenge of fusing what Wilfrid Sellars called the *manifest* and *scientific images* of a human being.[3] Central to the manifest image is the conception of a human being as a *person*: a free, rational, responsible agent who makes choices and acts for reasons. By contrast, the conception of a human being delivered by the scientific image is that of a complex bio-chemical system whose behavior conforms to causal laws. The challenge of fusing the images emerges when we note that the descriptions and explanations of human behavior that neuroscience is poised to provide appear to make no reference to rationality, responsibility, freedom, or choice, so even if we are convinced that each image is an adequate image of a human being, we seem to be in the dark about how the images are *related*. Put another way, there seems to be a conceptual gap between the terms belonging to the manifest image and the terms belonging to the scientific image, and fusing the images requires closing this gap. Attempts to bring neuroscience to bear on legal theory and practice will founder until we resolve this issue, for while neuroscience is situated within the scientific image, our legal system clearly has its home in the manifest image.

To get a better sense of what fusion requires and the challenge involved in meeting that requirement, it is useful to consider positions that *fail* to provide fusion. One such position traces back to the writings of René Descartes. Hoping to find a stopping point to the rising mechanical philosophy of the seventeenth century, Descartes famously argued that no purely physical system could think, reason, or be conscious. The conceptual gap entails a metaphysical gap, and thus was born mind-body dualism. According to dualism, there are facts about human beings—facts involving our status as rational, conscious, responsible agents—that cannot be fully accounted for by our physical nature, and so while a human being may be made up in part from physical ingredients, we must acknowledge nonphysical ingredients as well. In bifurcating human beings into physical and nonphysical ingredients, however, dualism leaves us with *fission*, not fusion.

Another position that fails to provide fusion rests on a fundamental agreement with dualism: the scientific image of human beings cannot be reconciled with the conception of humans provided by the manifest image. Instead of drawing the dualistic conclusion, however, this position encourages us to *abandon* all or many of the commitments belonging to the manifest image. Thus we hear that free will is an illusion,[4] as is any conception of responsibility that seems to require free will, including the conception that says criminals *deserve* punishment.[5] This position provides neither fission nor fusion, but rather *elimination* of core commitments of the manifest image.

I suspect that a lot of the anxiety generated by the emerging field of neurolaw comes from the belief both that fission and elimination are the only options and that neuroscience makes fission increasingly difficult to accept. Consider the widely discussed case of Charles Whitman. Whitman was an Eagle Scout, ex-marine, and engineering student who, on August 1, 1966, went on a shooting spree at the University of Texas, killing 13 people and wounding 32 others before being shot to death by police. Earlier that morning Whitman had murdered both his mother and wife, and in a suicide note composed the night before the murders, Whitman confessed to being a recent victim of many unusual and irrational thoughts. An autopsy revealed a large tumor in Whitman's brain which had impacted the amygdala, a structure of the brain implicated in regulating emotions, including fear and aggression. When we read about Whitman's case, it is hard to resist thinking *his tumor made him do it*. From here, it seems like a short step to the conclusion that *he is not responsible for his crimes*. But now we are faced with a very slippery slope: if we grant that it was the condition of Whitman's brain that made him kill, then shouldn't we admit that all human behavior has a biological basis (goodbye fission)? But if we admit that all human behavior has a biological basis, then shouldn't we conclude that no one is responsible for his or her actions (goodbye manifest image)?

Perhaps this anxiety could be relieved if we found an alternative to fission and elimination. This is what fusion hopes to provide. Fusion preserves the integrity of both images without bifurcating human beings into physical and nonphysical ingredients. Put another way, fusion allows that a purely physical system can be a free, rational, responsible agent. To those who seek to fuse the images, the issue of whether Whitman is responsible for his crimes does not turn simply on whether his behavior had a biological basis, but rather on whether his particular biology enabled the capacities necessary for free, rational, and responsible agency. In this way, fusion appears to provide a principled way to avoid the slippery slope.

The challenge for the defenders of fusion, of course, is to show *how* the images can be fused. There are two main obstacles to showing this. To help demonstrate the nature of the first obstacle, imagine that you are provided with a complete and accurate neuroscientific description of what is going on in some person's brain at some particular moment, say Noah's brain at 3:00 p.m. EST on November 20, 2015. Such a description might include terms like "axon," "dendrite," "action potential," "synapse," and "neurotransmitter," along with a host of other terms drawn from chemistry and physics. Now imagine being asked to conclude, based on this description alone, whether at that moment Noah was *thinking about the vacation he took to Florida in 2014*. Could you do it? There are powerful reasons for thinking that the answer is no. Though writing centuries before modern neuroscience, Gottfried Leibniz captured the difficulty in a vivid thought experiment:

> Moreover, we must confess that the *perception*, and what depends on it, is *inexplicable in terms of mechanical reasons*, that is, through shapes and motions. If we imagine that there is machine whose structure makes it think, sense, and have perceptions, we could conceive it enlarged, keeping the same proportions, so that we could enter into it, as one enters into a mill. Assuming that, when inspecting its interior, we will only find parts that push one another, and we will never find anything to explain a perception.[6]

The problem Leibniz anticipates—the problem of "Leibniz's Gap"[7]—is that the vocabulary employed in modern neuroscience does not seem conceptually linked to the vocabulary of "thought," "feeling," and "perception"—what we can call our *psychological* vocabulary. But without such a link, the neuroscientific description of Noah's brain provides us with no basis for concluding anything about what Noah is thinking. Such a basis is required, however, if we want to *show how* a purely physical system can also be a thinking, feeling, perceiving system. Overcoming the first obstacle thus requires us to provide the conceptual link that Leibniz claims is missing.

To help demonstrate the nature of the second obstacle, consider the picture of the universe inspired by Isaac Newton. According to Newton, the physical universe is governed by a handful of universal, exceptionless laws, i.e., laws that apply to all things, at all times, and at all places. Imagine that these laws can be expressed as mathematical equations. If we knew the complete state of the universe at any particular moment in time, we could plug that information into our equations and obtain a unique result. Here, the result would be a complete description of the state of the universe at the next moment in time. The critical feature is that the result is unique: given the complete state of the universe at any particular moment in time, plus the laws of nature, there is only one possible next state of the universe. That's determinism. Now consider whether there is free will in such a universe. On the face of it, the answer appears to be no. After all, it seems just obvious that part of what it means to say that I did something of my own free will is that I could have done otherwise than I did. But if I could have done otherwise, then there was more than one possible way for the universe to turn out. Free will is thus incompatible with determinism, so if the scientific image presents us with a deterministic physical universe, then either dualism is true and we must embrace fission or there is no free will and we must embrace elimination. Overcoming this obstacle thus requires showing that, contrary to first appearances, free will is possible in a deterministic universe.[8]

Minimally, fusion requires showing the compatibility of the images. But this is not enough. To see what more is needed, consider what Michael Pardo and Dennis Patterson say about a person stopping her car at a red light:

> Do the light waves from the lamp "cause" the pressure on the brake pedal? Surely not in the way the bowling ball causes the pins to fall over. It is true that the traffic light "causes" us to stop the car. But the "cause" for stopping the car cannot be explained solely by a physical process. By itself, the red light does not "cause" us to stop … rather, we stop because of the status of the light in an important social convention. … We are neither bowling balls nor pins. We have a choice. It is this choice that is the ground of responsibility.[9]

Note the claim that the cause for stopping the car cannot be explained solely by a physical process. If the cause for stopping the car is that which explains the stopping, then it looks like there will be facts about human beings—facts about our stopping cars, for example—that cannot be explained in terms of our physical nature. But if that is correct, then it seems that whatever we need to add to explain the car stopping will involve something nonphysical. Moreover, since what is missing appears to have something to do with choice, it is difficult to avoid the conclusion that choice is inexplicable on physical grounds—precisely the kind of conclusion drawn by those who embrace fission. But Pardo and Patterson want to preserve choice and responsibility while rejecting dualism, so they would have to deny that the explanation in mind introduces nonphysical entities or events. Their point seems to be that there is a kind of explanation at work here that is distinct from, and irreducible to, explanations that invoke causal relations between physical events, but this explanatory dualism does not entail metaphysical dualism. The problem with this tactic is that, while it is clearly in the spirit of fusion, it leaves the *relationship* between explanations couched in terms of the manifest image (e.g., choice and responsibility) and explanations couched in terms of the scientific image (e.g., causal relations between physical events) utterly mysterious. The burden for those who would hope to fuse the images is to show how the very same event—the stopping of the car—can be explained both in terms drawn from the manifest image and in terms drawn from the scientific image. Showing this requires providing ways to link the two vocabularies together, for without such a link, we will have no idea why an event explained in terms of causal relations between physical events can also be correctly explained in terms of agency and choice.

At the outset I said that attempts to bring neuroscience to bear on legal theory and practice will founder until the issue of whether we can fuse the manifest and scientific images of human beings is resolved, and the major task of this book is to show that they can be fused. Showing this requires closely examining and clarifying the key commitments of the manifest image—free will, rationality, and responsibility—and demonstrating that those commitments imply neither dualism nor indeterminism. Chapters 2–6 are dedicated to this task. The task begins by analyzing some recent attempts to draw popular attention to the issues underlying neurolaw. As I argue in Chapter 2, such attempts often betray subtle confusions or rest on problematic assumptions concerning free will, responsibility, and the relationship between mind and body. Identifying these confusions and assumptions will allow us to achieve a better understanding both of what the underlying philosophical issues facing neurolaw are and what the space of possible positions one can take on these issues looks like. Having articulated the relevant issues and the space of possible positions on these issues, I then turn to defending a position in that space. My point of departure is the recognition that the concepts of free will, responsibility, and rationality are intimately connected. We have encountered one aspect of this connection already, in the thought that responsibility requires free will. But as Stephen Morse points out, for the law, the issue of responsibility turns largely on the issue of rationality.[10] The account I give of free will, responsibility, and rationality aims to respect and illuminate these connections. The account starts in Chapter 3, where I invite us to take seriously the literal meaning of 'responsible': to be responsible is to be able to respond, in the sense of being able to modify one's decision-making procedures in ways appropriate to one's goals. Building on this conception of responsibility, Chapter 4 develops and defends an account of free will that identifies free will as responsiveness to rationality, in the sense that an ideal of rationality functions as a kind of focal point for adjustments in decision-making.[11] What emerges is an account of free will that is intimately linked with the notion of autonomy: an autonomous individual is self-ruling and *self-regulating*. The regulation concerns both the understanding, in the sense of tending toward increased knowledge, and the will, in the sense of tending to conform one's will to what one knows. On this account, the true threat to free will is not determinism, but rather intransigent forms of either ignorance or failures to coordinate understanding and will. I argue that this way of conceiving the threats to free will provides a fruitful framework for thinking about the challenges to law presented by insanity, addiction, psychopathy, and mental illness more generally.

The next challenge is to show how a purely physical system can have the kind of free will just characterized, and this challenge is addressed in Chapters 5 and 6. Key to showing this is the explanatory strategy of functional analysis, which is the attempt to explain the properties of complex systems by the analysis of a systemic property into organized interaction among other simpler systemic properties or properties of component subsystems. Reconciling this conception of a person with the image of humans provided by neuroscience first requires providing a functional analysis of this capacity to respond; in short, what kind of system design would exhibit the responsiveness that is needed for free will? Drawing on work done over the last 50 years in philosophy of mind and cognitive science, I show that this capacity can be explained in terms of the interaction of cognitive, affective, and volitional capacities, capacities that can in large part be understood in terms of representation, information, and computation. These capacities are characterized in terms of the contributions they make to a containing system, i.e., they are characterized in terms of their functions and not their intrinsic constitution. When we identify something functionally—a mousetrap, a gene, a legislature—we identify it in terms of what it does. Many biological terms have both a functional and an anatomical sense—an artificial heart is a heart by function but is not an anatomical heart—and cognitive neuroscience was conceived when "brain" became a functional term as well as an anatomical one. As I show, it is precisely this functional characterization that allows us to bridge the conceptual gap between our psychological and neurobiological vocabularies.

The discussion to this point has focused on long-standing philosophical questions concerning free will, responsibility, and the relationship between mind and body, but while the field of neurolaw compels us to face these crucial questions, other matters also deserve our attention. To drive this point home, imagine you are a juror faced with the following situation. There is a person, Nickey, on trial for first-degree murder. Nickey has pleaded not guilty by reason of insanity, and at one point during the court proceedings Nickey's defense attorney introduces the image below into evidence (Figure 1.1). Nickey's defense attorney then turns to the jury and declares, "Ladies and gentlemen, this image shows that my client was insane at the time of the murder and thus should be found not guilty." What would, or should, you make of this declaration? To answer this question, we need to identify, in a preliminary way at least, the crucial factors in play here. We have already identified one of them: How do we bridge the conceptual gap between our

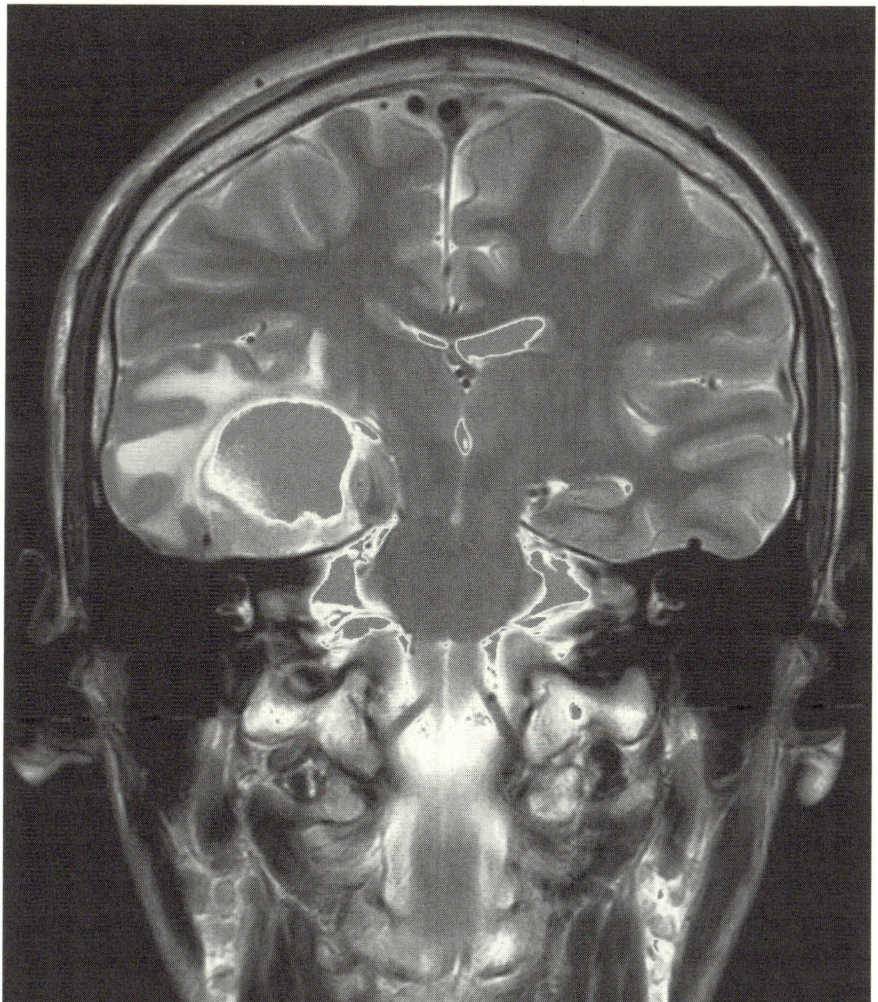

Figure 1.1 A representative brain scan generated by MRI. iStock.com/akesak.

neuroscientific vocabulary and our psychological vocabulary? But Nickey's case is more complicated than Noah's, in at least two ways. First, the issue raised in Noah's case doesn't concern how we come to *know* about the state of his brain; rather, it is about the conceptual connection between two different vocabularies. Even if we figured this out, it wouldn't help in Nickey's case unless we actually knew what state her brain was in at the time of the murder. In other words, Nickey's case introduces a host of issues about the *evidence* we have for the neuroscientific descriptions we might give of Nickey's brain—and the psychological descriptions we might be able to infer from them. A second complexity is that the issue of whether Nickey is sane or insane, guilty or not guilty, is not something that can be settled by facts about her brain or psychology alone. "Sanity" and "guilt" express *legal* concepts that have important implications for how we treat people (e.g., whether to incarcerate or let go), and so while these concepts are characterized partly in terms of our ordinary psychological concepts, they are not identical to them.

We can thus think of the core issues confronting neurolaw as falling into four broad categories, what I will call *metaphysical*, *moral*, *social*, and *evidential*. The metaphysical issues address the *nature* of human beings—what kind of stuff are we made of and how does this stuff work?—and as we have already touched on, of foremost concern are the mind-body problem and the possibility of free will. But while free will is certainly a traditional concern of metaphysics, understanding the connection between free will and responsibility requires us to consider both in relation to the practice and purpose of moral evaluation.[12] The metaphysical and moral issues are the primary focus of Chapters 2–6. The social issues address the aims of our criminal justice system, including the practice of holding people criminally responsible and the underlying rationale for adopting the legal definitions and distinctions that inform judgments of legal responsibility. Consider, for example, the concept of *insanity*. In the history of American jurisprudence there have been different, albeit related, ways to define or test insanity, most notably some variation or other of the M'Naghten Rule.[13] While it is plausible to think of each definition or test as reflecting our attempt to enshrine in law the manifest image of persons, the variability in definition reveals both that practical attempts to do this are not straightforward and that our legal definitions and distinctions are subject to negotiation, revision, and refinement. When we closely examine the definitions or tests currently employed by our criminal justice system, there are a number of questions we should ask: What do these definitions assume about human beings and what do these assumptions imply about choice, responsibility, guilt, and punishment? Given our current state of knowledge, are the assumptions reasonable? If we have good reasons for challenging these assumptions, how might we revise our definitions or tests to better reflect what we know about human beings?

Chapter 7 explores the significance of free will for being a member of a society like ours, a society in which members hold each other (and themselves) responsible for their actions and are entitled to praise or blame, reward or punish, depending on the actions so performed. The account of free will developed in this book implies that there can be degrees of free will, that is, variability in the extent to which someone can respond in the way distinctive of free persons. This variability is reflected in the variability of norms we apply to behavior, including age restrictions for driving, voting, and drinking, and in a host of other formal and informal ways. But we do largely assume that the members of this community are responsive. I argue that this assumption provides the principle rationale for judgments of criminal responsibility, and my argument is based on the content of existing law itself, including the Supreme Court's landmark ruling in *Atkins v. Virginia.* The connection between free will and criminal responsibility is further explored through an examination of various legal definitions of insanity and the nature of addiction and psychopathy. A presumption that provides the underlying rationale for many legal definitions of insanity is that knowledge of the nature of an act performed is necessary and sufficient for responsibility. Because insanity is a form of ignorance, it constitutes one way in which human free will—and thus responsibility—can be compromised. But the *presence* of knowledge is not enough to be free and responsible, for neuroscientific results provide compelling evidence that there are people for whom knowledge of what they are doing—including whether it is "right" or "wrong"—can have no effect on decision-making.[14] I argue that much psychopathic behavior can be explained this way; psychopaths typically appreciate the rationale for the very norms they flout, but this knowledge is nothing to them in the sense that it isn't action guiding. I discuss how we should think about criminal responsibility in light of these points, as well as implications for how we should treat many defendants suffering from mental retardation or other forms of cognitive or emotional impairment.

The evidential issues concern the methods or procedures by which we come to *know* about the biological and psychological states of people. For neurolaw, the central concerns revolve around the instruments we use to gather evidence about brains and what we may or may not reasonably infer from such evidence. Returning to the example of Nickey, what does the image purport to be an image of? How was the image produced? How reliable is the information provided by the image? Achieving clarity here will require an examination of both the various technologies used to observe brains—including positron emission tomography (PET) and functional magnetic resonance imaging (fMRI)—and the underlying theoretical commitments that inform our interpretations of evidence and the inferences we make from such evidence. It will also require us to confront complications that arise for forensic neuroscience, or the attempt to use the tools of neuroscience to investigate crime. For example, if the image was produced after the commission of a crime, what can we reasonably infer from this image about Nickey's brain state at the time of the crime? If the "machinery" of free will is located in our biology, then neuroscience has the potential to reveal the details of that machinery and provide insight into the conditions that disrupt it. Chapter 8 focuses on what neuroscience can potentially reveal about this machinery and addresses some of the complexities involved in realizing this potential.

Chapter 9 considers the overall adequacy of our current legal system in light of the considerations of the previous eight chapters. I argue that the basic rationale of our legal system is sound and not threatened by neuroscience as such. However, I also argue that our legal system has a number of serious shortcomings, and I indicate ways the legal system can be reformed to address them. The principle shortcoming is that our legal system does not adequately reflect the varying degrees of free will found among members of our population. Rather, the tendency is to find someone either insane and *not* criminally responsible or sane and *fully* criminally responsible. In effect, it tends to treat knowledge of the criminality of one's actions as the only condition relevant to responsibility. While there have been recent attempts to counteract this tendency, I argue that many of these attempts are deeply flawed and suggest ways we might reform our criminal justice system while avoiding these flaws. I also discuss the ways in which challenges to our legal system will have to be addressed at a much larger scale, including the problem of linking conformance to law with rationality. The problem, in short, is this: if a person has a good reason to think that she should *not* obey the law, then this person has a good reason *not* to take legal responsibility for her crimes. A reformed legal system will make sense only if the link between rationality and conformance to law is present; otherwise, our criminal justice system will be required to *hold* people legally responsible for actions they have no good reason to *take* legal responsibility for.

While legal scholars and neuroscientists have written extensively on many of the topics addressed in this book, explicit discussion of the relevant philosophical issues is often incomplete, misleading, or simply nonexistent in those works. On the other hand, philosophers often address the pertinent philosophical issues in a way that does not engage directly with the concerns of neuroscientists or legal scholars. The aim of this book is to provide such an engagement. This engagement is needed because our current criminal justice system has significant flaws, flaws that often result in the serious mistreatment of human beings. Many of these flaws are rooted in problematic conceptions of free will, rationality, and responsibility, and while identifying and improving these conceptions is not an easy task, it is an important one. Moreover, because the issues at stake here clearly impact the law—and public policy more generally—it is crucial that citizens of a democratic society like ours be able to discuss these issues in an informed way. I hope this book helps accomplish that task.

NOTES

1. http://www.lawneuro.org/.

2. Interest in the issues surrounding neurolaw is also reflected in a number of excellent anthologies that have appeared in recent years, including *The Future of Punishment*, ed. Thomas Nadelhoffer (New York: Oxford Series in Neuroscience, 2013); *Neuroscience and Legal Responsibility*, ed. Nicole Vincent (New York: Oxford Series in Neuroscience, 2013); *A Primer on Criminal Law and Neuroscience*, ed. Stephen Morse and Adina Roskies (New York: Oxford Series in Neuroscience, Law, and Philosophy, 2013).

3. Wilfrid Sellars, "Philosophy and the Scientific Image of Man," in *Science, Perception, and Reality* (Atascadero, CA: Ridgeview Publishing Company, 1963).

4. Daniel Wegner, *The Illusion of Conscious Will* (Cambridge, MA: MIT Press, 2002).

5. Joshua Greene and Jonathan Cohen, "For The Law, Neuroscience Changes Nothing and Everything," *Philosophical Transactions of the Royal Society B: Biological Sciences* 359 (2004): 1775–1785.

6. Gottfried Leibniz, *The Monadology*, in *Philosophical Essays*, trans. Roger Ariew and Daniel Garber (Indianapolis: Hacket, 1989), 215.

7. Robert Cummins, "'How Does It Work?' versus 'What Are the Laws?'" in *Explanation and Cognition*, ed. Frank C. Keil and Robert A. Wilson (Cambridge, MA: The MIT Press, 2000), 127.

8. Or a near enough deterministic universe. On the standard interpretation of quantum mechanics, quantum phenomena are indeterministic. However, the indeterminism at the quantum level is largely irrelevant to large-scale phenomena, including brains, which for practical purposes can be treated as deterministic systems. I return to this point in Chapter 2.

9. Michael Pardo and Dennis Patterson, *Minds, Brains, and Law: The Conceptual Foundations of Law and Neuroscience* (New York: Oxford University Press, 2013), Kindle edition, 40–41.

10. Stephen Morse, "New Neuroscience, Old Problems," in *Neuroscience and the Law: Brain, Mind, and the Scales of Justice*, ed. Brent Garland (New York: Dana Press, 2004), 157–198.

11. John Martin Fischer and Mark Ravizza (*Responsibility and Control: A Theory of Moral Responsibility* (New York: Cambridge University Press, 1998)) defend an account of free will according to which having free will amounts to being responsive to reasons, but while I think the positions developed in the present book complement Fischer and Ravizza's basic approach in a number of important ways, there are also notable differences in aim and focus, e.g., my account of free will emphasizes the link between free will and the role that the ideal of rationality plays in *explaining* the kind

of responsiveness that Fischer and Ravizza have in mind. Another difference concerns the respective weight we give to so-called "Frankfurt cases" in discussions of moral responsibility. I will have more to say about this over the next few chapters.

12. Gideon Yaffe, "Book Review: Responsibility and Control: A Theory of Moral Responsibility," *Erkenntnis* 53 (2000): 429–34.

13. The M'Naghten Rule is discussed in greater length in Chapter 7.

14. For an overview, see Robert Sapolsky, "The Frontal Cortex and the Criminal Justice System," *Philosophical Transactions of the Royal Society B: Biological Sciences. B* 359 (2004): 1787–1796.

Chapter 2

Fusion Confusion

In Plato's dialogue *Euthyphro*,[1] Socrates encounters a man, Euthyphro, who is about to try his own father for murder. Recognizing the serious nature of this matter, Socrates asks Euthyphro whether he is fully confident that he is doing the right thing. Euthyphro assures him that he is, explaining that piety requires him to prosecute his father. Socrates wants to know what piety is but believes himself ignorant on the matter, and because it would be reckless of Euthyphro to prosecute his father in the name of piety unless Euthyphro knows what piety is, Socrates pleads Euthyphro to teach him the nature of piety. Euthyphro agrees, but Socrates demonstrates that Euthyphro's attempts to define piety are riddled with ambiguity, confusion, and error. Euthyphro is left in a state of puzzlement, a state that the ancient Greeks called *aporia*. Despite this, at the end of the dialogue we learn that Euthyphro still intends to prosecute his father!

While aporia is a state of puzzlement that indicates *lack* of knowledge, it can play a critical role in achieving knowledge. Because Euthyphro is initially unaware of his own ignorance, he has no reason to undertake investigations that might get him closer to the truth. Like Euthyphro, we may believe that those appointed to prosecute, convict, and sentence defendants are largely justified in what they do and that support for these practices comes from widely shared beliefs about the nature of mind, rationality, free will, and responsibility. Because these beliefs provide our justification for these practices, anything that undermines our confidence in these beliefs should also undermine our confidence in the legitimacy of these practices. What are these beliefs? What reasons might we have for holding them? Are there plausible grounds for doubting them? My hope in exploring these questions is to make the seemingly obvious seem a lot less so—to induce aporia, in short. Our exploration will begin by examining some attempts to bring popular attention to neurolaw. The principle reason for starting with these attempts is that, because such attempts are aimed at a wide audience, each can be plausibly viewed as giving voice to familiar and widely shared beliefs about the nature of mind, rationality, free will, and responsibility that we are interested to explore.

NEUROLAW IN THE AGORA

The first attempt we'll consider comes from a *New York Times Magazine* article by Jeffrey Rosen. Rosen presents two contrasting views on the implications of neuroscience for the law. The first view is articulated by Harvard psychologist Joshua Greene:

> "To a neuroscientist, you are your brain; nothing causes your behavior other than the operations of your brain," Greene says. "If that's right, it radically changes the way we think about the law. The official line in the law is all that matters is whether you're rational, but you can have someone who is totally rational but whose strings are being pulled by something beyond his control." In other words, even someone who has the illusion of making a free and rational choice between soup and salad may be deluding himself, since the choice of salad over soup is ultimately predestined by forces hard-wired in his brain. Greene insists that this insight means that the criminal-justice system should abandon the idea of retribution—the idea that bad people should be punished because they have freely chosen to act immorally—which has been the focus of American criminal law since the 1970s, when rehabilitation went out of fashion. Instead, Greene says, the law should focus on deterring future harms.[2]

The second view is articulated by Stephen Morse, a professor of law and psychiatry at the University of Pennsylvania:

> "There's nothing new about the neuroscience ideas of responsibility; it's just another material, causal explanation of human behavior. … How is this different than the Chicago school of sociology, which tried to explain human behavior in terms of environment and social structures?" Morse agrees that our brains bring about our behavior—"I'm a thoroughgoing materialist, who believes that all mental and behavioral activity is the causal product of physical events in the brain"—but he disagrees that the law should excuse certain kinds of criminal conduct as a result. "It's a total non sequitur," he says. "So what if there's biological causation? Causation can't be an excuse for someone who believes that responsibility is possible. Since all behavior is caused, this would mean all behavior has to be excused." Still, Morse concedes that there are circumstances under which new discoveries from neuroscience could challenge the legal system at its core. "Suppose neuroscience could reveal that reason actually plays

> no role in determining human behavior," he suggests tantalizingly. "Suppose I could show you that your intentions and your reasons for your actions are post hoc rationalizations that somehow your brain generates to explain to you what your brain has already done without your conscious participation." If neuroscience could reveal us to be automatons in this respect, Morse is prepared to agree with Greene and Cohen that criminal law would have to abandon its current ideas about responsibility and seek other ways of protecting society.[3]

As a way into this dialectic, consider the example Rosen gives of choosing between salad and soup. Suppose you go with the salad. While you may believe that the choice of salad was free and rational, we are presented with an argument that is supposed to undermine this belief. Here's the argument:

Premise 1: If forces hard-wired in the brain ultimately predestine the choice of salad over soup, then the choice of salad over soup is not free and rational.
Premise 2: Forces hard-wired in the brain ultimately predestine the choice of salad over soup.
Conclusion: The choice of salad over soup is not free and rational.

The argument looks valid, so rejecting the conclusion requires rejecting at least one of the premises. Let's start with the first premise. The condition that is supposed to undermine free and rational choice concerns three salient features of the forces at work here. These forces are *hard-wired*, they are *in the brain*, and they *predestine choice*. If Premise 1 strikes you as true, ask yourself what contribution each of these features makes to the plausibility of the claim. Let's start with being hard-wired. The phrase "hard-wired" typically refers to something that is genetically determined, so if being hard-wired is crucial to the plausibility of the claim, we would expect the plausibility to be diminished at least somewhat if it turned out that these forces were not hard-wired. But suppose that the forces were not genetically determined but rather were determined in the first few months after your birth by some environmental condition, e.g., after exposure to lead-based paint. It is hard to see how that difference could matter to whether your choice was free and rational. In both scenarios the forces are determined and in both scenarios you have no control over the determining factors. What matters, it seems, is *that* factors beyond your control determined these forces, *not how* factors beyond your control determined these forces. This is the point Morse is getting at when he refers to the Chicago school of sociology.

Turning to the second feature, we can similarly test the relevance of *being in your brain* by considering whether being "in" some other kind of thing would affect the plausibility of the premise. For example, suppose Rosen had interviewed a Catholic theologian instead of a neuroscientist. Imagine that this theologian is named Father Rede, and assume Father Rede says exactly what Greene says, except that when Greene says "brain" Father Rede says "nonphysical soul." We'd get:

> "To a theologian, you are your non-physical soul; nothing causes your behavior other than the operations of your non-physical soul," Rede says. "If that's right, it radically changes the way we think about the law. The official line in the law is all that matters is whether you're rational, but you can have someone who is totally rational but whose strings are being pulled by something beyond his control." In other words, even someone who has the illusion of making a free and rational choice between soup and salad may be deluding himself, since the choice of salad over soup is ultimately predestined by forces hard-wired in his non-physical soul.

In broader terms, we are now wondering whether it matters to free and rational choice that the "part" of you that is relevant to choosing salad is physical or not. Remember that we are holding fixed the fact that what you choose is based on forces "in" your brain or "in" your nonphysical soul, and we are holding fixed that there were factors beyond your control that determined the presence of these forces, and we are holding fixed that these forces predestine the choice. As with the distinction between genetic and environmental factors, it is hard to see how the distinction between physical and nonphysical matters *per se* to the issue of free and rational choice. I suspect the reason it seems to matter is that when we think "physical" we think "determined" and simply assume that "not physical" implies "not determined." But it doesn't, as a parallel example shows. If Fido is a dog, then Fido is a mammal. But if Fido isn't a dog, it doesn't follow that Fido isn't a mammal (Fido could be a cat). However, even if "not physical" *is* taken to suggest "not determined," it is important to note that it is the "not determined" part implied by "not physical" that is doing all the work in our judgments about free and rational choice.

So what about the predestined part? Setting aside any theological connotations suggested by "predestined," to be predestined is simply to be determined, i.e., the forces that led to your choice of salad over soup determined the choice of salad. This feature of the forces does look relevant to the plausibility of Premise 1; as we will see, not being determined has seemed to many part of the very concept of free will, so if the forces relevant to choosing salad did *not* determine this choice, it would be hard to understand why the presence of such forces threatens free will.

In light of our examination of the features that makes Premise 1 plausible, we should revise that premise as follows:

Premise 1:* If there are factors that determine the existence of forces in you, forces that in turn determine the choice of salad over soup, then the choice of salad over soup is not free and rational.

As the revised premise should make clear, the threat to free will is supposed to result from the fact that the forces in play here are (i) determined by prior factors and (ii) determine choice. Minimally, then, salvaging free will seems to require both that prior factors not determine the forces involved and that these forces do not determine our choices.

Is Premise 1* true? If it is, does this matter to the law? Before addressing these questions directly, we need to consider whether the features of forces that supposedly threaten free choice also threaten rationality. It is a bit odd for Rosen to claim that the features of forces that supposedly threaten free choice also threaten rational choice, for Greene himself explicitly states that a person can be rational even if the forces at work have the features that supposedly threaten free choice. To see the compatibility, note first that we have said virtually nothing about the precise nature of these forces. All we have focused on to this point is whether the presence of these forces is determined by prior conditions and whether these forces themselves determine choice. So what kind of forces might be relevant to the choice of salad over soup? Suppose each of the following is true: (a) you want to be healthy and believe that the salad is the healthier option; (b) you want to save money and believe the salad is less expensive than the soup; (c) you prefer the taste of soup over salad. (a) and (b) provide reasons for choosing the salad, while (c) provides a reason for choosing the soup. A rational person is a person who acts in accordance with her reasons, so if you judge that the balance of reasons favors salad over soup and choose salad *because* the balance of reasons favors it, then you are acting rationally. Notice that this characterization of what it is to be rational says nothing about where the reasons come from or whether the reasons determine the choice. As far as rationality goes, your reasons could be both the result of prior, determining factors and constitute forces that determine your choice.

Greene and Morse each accept that a person can be rational even if determinism is true. Their disagreement is over whether questions of legal responsibility turn *only* on questions of rationality. Though Rosen does not discuss it, in an article written by Greene and his colleague Jonathan Cohen, they argue that our judgments about responsibility do not turn only on judgments of rationality. To demonstrate this, Greene and Cohen use a thought experiment about a man named Mr. Puppet. Mr. Puppet was created by a group of scientists and is now on trial for a crime he committed. The defense attorney calls the lead scientist to the stand, and the scientist testifies as follows:

> I designed him. I carefully selected every gene in his body and carefully scripted every significant event in his life so that he would become precisely what he is today. I selected his mother knowing that she would let him cry for hours and hours before picking him up. I carefully selected each of his relatives, teachers, friends, enemies, etc. and told them exactly what to say to him and how to treat

> him. Things generally went as planned, but not always. For example, the angry letters written to his dead father were not supposed to appear until he was fourteen, but by the end of his thirteenth year he had already written four of them. In retrospect I think this was because of a handful of substitutions I made to his eighth chromosome. At any rate, my plans for him succeeded, as they have for 95% of the people I've designed. I assure you that the accused deserves none of the credit.[4]

Greene and Cohen then ask, "What to do with Mr. Puppet?" Here is their answer:

> Insofar as we believe this testimony, we are inclined to think that Mr. Puppet cannot be held fully responsible for his crimes, if he can be held responsible for them at all. He is, perhaps, a man to be feared, and we would not want to return him to the streets. But given the fact that forces beyond his control played a dominant role in causing him to commit these crimes, it is hard to think of him as anything more than a pawn. But what does the law say about Mr. Puppet? The law asks whether or not he was rational at the time of his misdeeds, and as far as we know he was. For all we know, he is psychologically indistinguishable from the prototypical guilty criminal, and therefore fully responsible in the eyes of the law. But, not intuitively, this is not fair.[5]

Thus when Morse says that causation can't be an excuse for someone who believes that responsibility is possible, Greene and Cohen would reply that responsibility is *not* possible in a deterministic universe, we live in a deterministic universe, and thus we should abandon the belief that responsibility is possible.

While Rosen remains neutral on the dispute between Greene and Morse, David Eagleman clearly agrees with Greene. In "The Brain on Trial," Eagleman notes that "The legal system rests on the assumption that we are 'practical reasoners,' a term of art that presumes, at bottom, the existence of free will. The idea is that we use conscious deliberation when deciding how to act—that is, in the absence of external duress, we make free decisions."[6] To illustrate what Eagleman means by using conscious deliberation to decide how to act, let's return to the example of deciding between salad and soup. Considerations of health and budget favored choosing salad, while considerations of taste favored soup. You considered the reasons for and against each choice, judged that the balance of reasons favored the salad, and decided to order salad. Two features are critical to the example. First, you were consciously aware of the reasons for and against ordering each item, as well as the judgment about which item the balance of reasons favored ordering. If you "decided" to order salad but lacked awareness of any reason for ordering it or lacked awareness of why the balance of reasons favored it, then your decision would not be the result of conscious deliberation. Second, what you ended up deciding depended on the judgment you made about which item the balance of reasons favored. If it had turned out that your decision depended only on the outcome of a roll of dice, say, then the fact that the balance of reasons favored salad would have been only coincidentally related to your choice.

Before going any further, we should ask whether there is any significant difference between being a rational agent and being a conscious deliberator. Morse says that the law cares about whether a person is rational, and the illustration of conscious deliberation I just gave certainly looks like an illustration of rationality as well. The question is important because if determinism doesn't threaten rationality and being a rational agent and being a conscious deliberator are more or less the same thing, then determinism won't threaten conscious deliberation. We'll return to this point shortly.

Eagleman claims that this conception of the practical reasoner is "intuitive but problematic."[7] The conception is intuitive in that it seems obvious that much of our behavior results from conscious deliberation, but it is problematic in that much behavior can take place in absence of conscious deliberation. Eagleman discusses several examples of this, including Tourette's syndrome. People with Tourette's are prone to involuntary behavior, including verbal behavior, and Eagleman thinks this has important implications:

> We immediately learn two things from the Tourette's patient. First, actions can occur in the absence of free will. Second, the Tourette's patient has no *free won't*. He cannot use free will to override or control what subconscious parts of his brain have decided to do. What the lack of free will and the lack of free won't have in common is the lack of "free." Tourette's syndrome provides a case in which the underlying neural machinery does its thing, and we all agree that the person is not responsible.[8]

Eagleman also notes that surprisingly sophisticated behavior can take place in the absence of conscious deliberation and illustrates the point with the bizarre case of Kenneth Parks. On May 23, 1987, Parks stabbed his mother-in-law to death. At trial, Parks' defense attorney argued that Parks committed the murder while sleepwalking and thus was not conscious of what he was doing. The jury agreed, finding Parks not guilty. Eagleman thinks that Parks' case has important implications for free will, writing that the case "illustrates that high-level behaviors can take place in the absence of free will. Like your heartbeat, breathing, blinking, and swallowing, even your mental machinery can run on autopilot."[9]

At this point you may be thinking that, while important, the examples Eagleman gives are rare exceptions and that most of the behavior that the law concerns itself with is the result of conscious deliberation. Not so, according to Eagleman:

> The crux of the question is whether *all* of your actions are fundamentally on autopilot or whether some little bit of you is "free" to choose, independent of the rules of biology. This has always been the sticking point for philosophers and scientists alike. After all, there is no spot in the brain that is not densely interconnected with—and driven by—other brain parts. And that suggests that no part is independent and therefore "free." In modern science, it is difficult to find the gap into which to slip free will—the uncaused causer—because there seems to be no part of the machinery that does not follow in a causal relationship from the other parts. Free will *may* exist (it may simply be beyond our current science), but one thing seems clear: if free will *does* exist, it has little room in which to operate. It can at best be a small factor riding on top of vast neural networks shaped by genes and environment. In fact, free will may end up being so small that we eventually think about bad decision-making in the same way we think about any physical process, such as diabetes or lung disease.[10]

The first sentence implies that if there is no part of you that can choose independent of the rules of biology, all of your actions are on autopilot, i.e., all of your actions occur without conscious awareness. Sentences four and five basically say that modern science finds no part of you that can choose independent of the rules of biology. From this it follows that, according to modern science, all of your actions are on autopilot. If Eagleman is right that a person suffering from Tourette's is not responsible for her involuntary behavior and she is not responsible because her actions are on autopilot, then according to modern science no one is responsible for anything they do. And finally, if all actions are on autopilot, then modern science says that you are no different than Kenneth Parks!

If you find the final conclusion particularly absurd it's likely because Eagleman conflates several plausible, but distinct, senses of "autopilot." Typically, people who suffer from Tourette's are consciously aware of their behavior, e.g., a person who suffers from Tourette's is consciously aware that she is blurting out an inappropriate remark. Parks' case is different in that he was not even consciously aware of killing his mother-in-law. The sense in which a person suffering from Tourette's is on autopilot is that the relevant behavior is typically not subject to control by any conscious process of reflection and evaluation. But the phrase "control by any conscious process of reflection and evaluation" is slippery. Return for a moment to the salad example. In that case we imagined a process in which the person consciously considered reasons for and against ordering each item and was consciously aware of the judgment that the balance of reasons favored ordering salad. Does a similar process occur for most of the decisions we make? Probably not. This morning I ate a banana, as I do most mornings. I was consciously aware of eating the banana, and I would say that I decided to eat the banana. But at no point this morning did I consciously reflect on reasons for and against eating the banana. Nor did I consciously reflect on reasons for and against eating another fruit, or eating something other than a fruit, or eating nothing at all. Does that mean that I was on autopilot in the sense in which the behavior of a person suffering from Tourette's is on autopilot? No, for while the proximate cause of my eating a banana this morning was not a conscious process of reflection and evaluation, it is plausible to think either that I behaved the way I did today because of an earlier act of conscious reflection and evaluation, or that acts of conscious reflection and evaluation can matter to whether the behavior occurs. If, for example, someone asked me why I ate the banana—and ate the banana instead of an apple, or instead of something other than a fruit, or instead of nothing at all—I could give the reasons that explain my behavior. Alternatively, if I could not give any (good) reasons for my behavior, then it is not far-fetched to imagine that I might stop my daily intake of bananas. Moreover, unlike the person who is suffering from Tourette's, the sense in which I was on autopilot may be a virtue. Suppose there was a moment in the past in which I consciously deliberated about what to eat in the morning, and suppose I judged that the balance of reasons favored eating a banana. If I also judged that *every morning I should eat a banana*, then consciously deliberating over what to eat each morning would be a waste of time and energy.

Of course, we might wonder whether in the imagined situation my banana eating ever resulted from the judgment that the balance of reasons favored it, even if I underwent a process of conscious deliberation. To get at the worry here, consider a widely discussed experiment performed by Richard Nisbett and Timothy Wilson. Nisbett and Wilson placed four pairs of pantyhose on a store table and asked customers to state which pair they preferred and explain why they preferred it. Customers mostly preferred the pair located on the right side of the table, and customers would cite features like color and sheerness to explain their preference. In fact, the four pairs of pantyhose were identical in color and sheerness.[11] Two related consequences are suggested by this experiment. First, it seems that the location of the pantyhose was clearly relevant to preference even though customers did not cite location when explaining their preferences. Second, the felt need of customers to make sense of their preferences led them to make distinctions that simply were not there. So not only were customers unaware of factors relevant to preference, they would *fabricate* factors in order to make sense of their preferences.

Let's take stock. We began with the worry that all of our behavior is on autopilot, and imagined that this meant we are all like Kenneth Parks. We noted a distinction between Parks and those who suffer from Tourette's, and noted another, arguably more important distinction between the behavior of those who suffer from Tourette's and the "routine" behavior most of us exhibit, as illustrated by the example of eating a banana. We then called into question whether conscious deliberation was relevant to the banana example, if only in the sense that conscious deliberation partly explained the now "autopilot" behavior. If every case of conscious deliberation was like the pantyhose case, then Eagleman would be right: conscious deliberation actually plays no role in decision-making, and we are all like those suffering from Tourette's. This worry is the worry that Morse also expresses, recall. But now note two things: first, we do not yet have an argument for the conclusion that all conscious deliberation is like the pantyhose case. Second, the pantyhose case is completely irrelevant to Eagleman's argument. I'll focus on the second point here. The pantyhose case, if generalized, would show that the reasons we give (or could give) for our decisions are actually irrelevant to what we do. The "real" causes operate in a way that is insensitive to conscious reflection and evaluation and could not constitute reasons, in the sense that they would not make the behavior rational. But if this threatens free will, the threat has nothing to do with whether the causes of behavior operate independent of the rules of biology. At no point did we need to assume that "behavior that results from conscious deliberation" entails "behavior that results from something independent of the rules of biology." Put another way, the threat had to do with the *kind* of causes operative—did (possibly) conscious processes of reflection and evaluation cause the behavior or not?—not *whether* the behavior was caused.

Let's reconstruct Eagleman's argument to help clarify the issues in play here. The argument appears to be this:

Premise 1: A choice is made out of one's own free will only if the choice results from one's conscious deliberation.
Premise 2: None of the factors relevant to choice operate independent of the rules of biology.
Conclusion: Therefore, no choices are made out of one's own free will.

As stated, the conclusion does not follow from the premises. To get a valid argument, we need another premise. This will do:

Premise 3: If a choice results from one's conscious deliberations, then there are factors relevant to choice that operates independent of the rules of biology.

Eagleman claims that the legal system rests on Premise 1, and the support for Premise 2 comes from modern neuroscience. However, even if we grant that both premises are uncontroversial, we've seen that the ominous conclusion follows only if we accept Premise 3. Should we accept it? Before answering this question, there are two points worth noting. The first is that if we accept Premise 3, then Eagleman's extended discussion of Tourette's and the case of

Kenneth Parks is completely irrelevant. The reason is this: the neuroscientific evidence that supports Premise 2 does not essentially depend on the existence of syndromes like Tourette's or cases like the one involving Parks. That is, even if no one had Tourette's or somnambulism, we'd still have compelling evidence that none of the factors relevant to choice operate independent of the rules of biology. So why bother discussing these cases? As I will argue, these cases are relevant to the question of the extent to which conscious deliberation matters to choice, but their relevance has nothing to do with the issue of whether there are factors relevant to choice that exist independent of the rules of biology. Put another way, the distinction between choices that result from conscious deliberation and those that do not is a distinction to be found *within* biology. If that's right, then I think we should say that Eagleman has simply misidentified what is important about the cases he mentions. This leads to the second point. Unlike the first two premises, Premise 3 is highly controversial and in need of argument. Instead of considering challenges to this premise or alternative ways to conceive of the relationship between conscious deliberation and biology, however, Eagleman simply assumes that Premise 3 (or something close to it) is true. Let's now take a closer look at this premise. We will have a good reason to accept it if we can show that there are properties of conscious deliberation that are excluded by the rules of biology. Eagleman's worry about whether all of our actions are on autopilot suggests that it is the conscious part of conscious deliberation that is excluded by the rules of biology, but perhaps it is the phenomenon of deliberation itself that generates the tension. Though Eagleman himself does not explicitly discuss the issue, some philosophers have argued that deliberation would be impossible if determinism were true. Consider the salad example again. The aim of my deliberation was to decide which item to order, to settle the matter one way or another. But this makes sense only if I believe that it is genuinely possible for me to order salad *and* it is genuinely possible for me to order soup. But as Richard Taylor notes, "If I believe that what I am going to do has been rendered inevitable by conditions already existing … then I cannot decide whether to do it or not, for there is simply nothing left to decide."[12] If determinism is true, then what I am going to do has been rendered inevitable by conditions already existing: if I order salad, ordering salad was the only genuine possibility. Thus, if determinism is true, then I do not decide anything and deliberation is impossible.

There are thus two ways in which we might see conscious deliberation threatened. The first says that much of the behavior that we intuitively think is the result of conscious deliberation is in fact the result of something else. The second says that the rules of biology render deliberation impossible. Eagleman initially frames the threat in the first way but ends up framing it in the second, and while each may constitute a genuine threat, it is crucial not to confuse them. I mentioned earlier that if the rules of biology pose a threat to free will, then Eagleman's discussion of Tourette's and Kenneth Parks is irrelevant. The reason should be clear. If all behavior is rooted in biology and the rules of biology rule out the very possibility of deliberation, then there can be no interesting question about the extent to which behavior is the result of conscious deliberation; *none of it could be*. This question can be interesting only if we accept that the rules of biology do not exclude conscious deliberation per se. We can thus say that the two threats to conscious deliberation have an importantly different character. The first threat does not turn on the assumption that determinism per se rules out conscious deliberation, i.e., it grants that behavior can be determined *and* the result of conscious deliberation. Rather, the threat is based on what science shows (or may come to show) about the *actual* sources of human behavior. But even here, the threat isn't that some, or even a lot, of behavior is not the result of conscious deliberation. The threat is that a lot of behavior *that we typically think is the result of conscious deliberation*—and thus behavior for which a person can be held responsible—is actually not. Whether such a threat exists, and if so, its extent, is something to be settled by empirical investigation.

By contrast, the second threat is more "philosophical" than "empirical." It says that reflection on the concepts of determinism, free will, and conscious deliberation reveals that conscious deliberation requires a kind of free will that determinism rules out, so the truth of determinism is sufficient to rule out conscious deliberation. The crucial claim here is the claim that determinism and free will are incompatible. Let's take a closer look at this claim.

NEUROLAW IN THE LYCEUM

The thesis that determinism rules out free will is the thesis of incompatibilism:

(I): If determinism is true, then there is no free will.

Note that (I) does not assert that determinism is true. Nor does it assert that there is no free will. Rather, it asserts that determinism and free will *exclude* each other. Suppose we accept (I). There are two ways we might reason from it. The first way would be to accept determinism and conclude that there is no free will. Those who reason this way are often called "hard determinists." The second way would be to accept free will and conclude that some choices are not determined. Those who reason this way are often called "libertarians" (no relation to the political philosophy).

Because many philosophers reject (I), we need to consider how an incompatibilist might support this thesis. Let us understand the claim "there is free will" as equivalent to "some choices are made of one's own free will." We say *some* choices because accepting free will does not require accepting that all choices are free. Strictly speaking, all that is required for "there is free will" to be true is that at least one agent makes at least one choice of her own free will. With this in mind, we might argue for (I) as follows. Letting "S" stand for any agent, "X" stand for any action, and "t" stand for any time

Premise 1: If S chose to do X, at time t, of S's own free will, then S could have chosen otherwise than X at time t.
Premise 2: If S could have chosen otherwise than X at time t, then more than one future was possible at t.
Premise 3: If determinism is true, then it is not the case that more than one future was possible at t.
Conclusion 1: Therefore, if determinism is true, then it is not the case that S could have chosen otherwise than X at time t.
Conclusion 2: Therefore, if determinism is true, then it is not the case that S chose to do X, at time t, of S's own free will.

Because the argument applies to any agent, any action, and any time, it follows that if determinism is true, then there are no instances in which a choice is made of one's own free will, i.e., there is no free will—precisely what (I) says.

This is a powerful argument, and it may make (I) seem obvious. But many philosophers have not been persuaded by it. These philosophers—called compatibilists—deny (I) and think the above argument is flawed. Recall that both libertarians and hard determinists accept (I). In rejecting (I), compatibilists thus disagree with both libertarians and hard determinists.

So what is the flaw in the argument, according to compatibilists? The flaw is not in the reasoning itself. Compatibilists agree that if all the premises are true, the conclusion has to be true as well. Rather, compatibilists think that one of the premises is false. Which one? An historically influential form of compatibilism—what I will call traditional compatibilism—urges us to accept both Premise 1 and Premise 3.[13] According to traditional compatibilism, the disagreement isn't over whether "free" implies "could have chosen otherwise" (as captured by Premise 1), nor is it over the definition of determinism and what is implied by determinism (as captured by Premise 3). Rather, the problem is that Premise 2 provides an incorrect analysis of the phrase "could have chosen otherwise." To see why the traditional compatibilist thinks this, consider the salad example again. In the situation I described, my choice of salad looked like a choice made of my own free will. If it was, then I could have chosen otherwise; I could have chosen soup, for example. Part of what is involved in my believing that the choice of soup was possible is that I can coherently imagine a circumstance in which the choice of soup was actual. If I could not even conceive of a world in which the choice of soup was actual, I would have no grounds for thinking the choice to be genuinely possible. So what am I imagining when I am imagining a world in which I choose soup? Recall that in the actual world I chose salad, and I made that choice on the grounds that the balance of reasons favored salad. When I imagine a world in which I choose soup, am I (i) imagining a world in which the balance of reasons favors salad, or am I (ii) imagining a world in which the balance of reasons favors soup? Traditional compatibilists think the answer is obviously (ii), for if I answer (i), then what I am imagining is a world in which *my choice is unconnected to my reasons for acting*. Picture the scenario provided in (i). I tell the waiter that I'll have the soup even though I believe salad is the better option. From my very own perspective my choice makes no sense! This is *not* what we mean by "could have chosen otherwise," says the traditional compatibilist. What we mean is that I *would have* chosen differently if my reasons had been different, e.g., if the balance of reasons had favored ordering soup I would have ordered soup. This reading of "could have chosen otherwise" captures the important way in which our choices reflect our reasons for acting, and it is precisely this fact about my choice of salad that makes the choice free.

So what does this have to do with compatibilism and incompatibilism? Well, if the world I imagine when I imagine choosing soup is also a world in which the balance of reasons favors soup, then I am *not* imagining choosing soup in the very circumstance in which I actually chose salad. The complete state of the universe at the time I chose salad included the fact that the balance of reasons favored salad. In those circumstances, choosing soup is not possible. But when I imagine choosing soup, I am imagining a slightly different situation—the situation in which the balance of reasons favors soup. But this is perfectly consistent with determinism. Determinism does not say that if an event happens, it happens regardless of what the world was like prior to the event. Birds are chirping outside of my window right now. If the Moon had crashed into the Earth one week ago, there would be no birds chirping outside my window right now.

The above line of reasoning has convinced many to embrace traditional compatibilism, but others, like Kant, have viewed this as a "wretched subterfuge."[14] The worry can be brought out clearly when we think about the alleged connection between free will and *responsibility*. Let's return to Mr. Puppet and suppose for the moment that the only consideration relevant to his free will and responsibility is whether he could have chosen otherwise than to commit the crime. Let's further suppose that he would have chosen otherwise if his reasons for action had been different. Does this make him responsible for the choice? If you don't think so, I suspect it is because, though he would have chosen otherwise if his reasons for action had been different, his reasons for action could not have been different. Or rather, when we consider the conditions under which it is true that his reasons for action could have been different—they would have been different if something else had been different—we eventually reach a point where whether the conditions under which his reasons for action would have been different obtain have nothing to do *with him*. We can make this clear with the salad example. I would have chosen soup if the balance of reasons had favored soup. Could the balance of reasons have favored soup? If so, are the conditions under which the balance of reasons *would* have favored soup (remember, "could" chases "would") conditions for which I am responsible? One of the reasons that favored ordering salad was my desire to be healthy and my belief that the salad was the healthier option. Could I have not desired to be healthy? Suppose at some point in my youth I realized that I wanted to live a long life and believed that being healthy was necessary to living a long life. In that case it may be true that I would not have wanted to be healthy if I hadn't wanted to live a long life. Putting aside the consideration of cost, let's stipulate that the situation in which I don't want to be healthy is a situation in which the balance of reasons favors soup. Because the fact that I wanted to be healthy in part resulted from my wanting to live a long life, we might say that the conditions under which the balance of reasons would have favored soup are conditions that depend on me, e.g., they depend on whether I want to live a long life—and thus I am responsible for the fact that the balance of reasons favored salad and thus responsible for the choice as well. OK, but are you responsible for wanting to live a long life? Hopefully you see where this is heading. Maybe I am responsible for wanting to live a long life, but when we go back far enough we will eventually reach some point where *my* responsibility bottoms out, i.e., whether the circumstances in which things would have been different obtain don't depend on me. If determinism is true, then there are conditions that don't depend on me and for which I am not responsible—conditions existing before my birth, for example—such that, given those conditions, the balance of reasons favoring salad (and thus my choice of salad) is the only possible outcome.[15] It seems that I am not responsible for my choice of salad even though it is true that I would have chosen soup if the balance of reasons had favored soup. The truth of that counterfactual thus appears irrelevant to responsibility, so if responsibility for a choice requires that the agent could have chosen otherwise, the traditional compatibilist gives the wrong analysis of "could have chosen otherwise."

Does the incompatibilist who affirms free will, i.e., the libertarian, fare any better here? Troubles for libertarian free will arise if we accept both (a) *you are your brain*, and (b) *brains are deterministic physical systems*, so the libertarian needs to deny at least one of these claims. Because the rise of mechanical philosophy made rejecting (b) increasingly difficult, some, including Descartes, opted to reject (a). However, it should be fairly evident that rejecting (a) will help here only if we also reject (b). Here's why. It is a datum of common sense that our choices affect physical behavior. For example, whether I pick up the fork or the spoon depends on whether I choose the salad or soup. But if my choice of salad is not determined, then either the behavior of my body is not determined or free choice is irrelevant to behavior. To see this, consider the complete state of the physical world at the time I freely chose the salad. If my choice was free, then the choice to have soup was possible at that time and in those exact conditions. If free choice is relevant to physical behavior, then my physical behavior would have been different had I chosen soup. But if my physical behavior was determined but my choice was not, then I would have ended up picking up the fork even if I had chosen soup. Thus either my physical behavior is not determined or free choice is irrelevant to behavior.

While rejecting (a) can help salvage libertarian free will only if we also reject (b), rejecting (b) would not require us to reject (a). As far as libertarianism is concerned, it does not matter whether the indeterminism required for free choice resides in the nonphysical soul or the physical brain, and many contemporary libertarians, including Robert Kane, reject (b) but accept (a).[16] Kane's rejection of (b) is based on the standard reading of quantum mechanics according to which the behavior of matter at very small scales is not determined. However, it is not clear whether quantum indeterminism can really help the libertarian. Greene and Eagleman write as if determinism is true, but it is not as though they are ignorant of quantum mechanics or reject the standard, indeterministic reading of it. Rather, their determinism is based on the plausible assumption that the effects of quantum indeterminism are largely cancelled out when we consider larger scales of matter, including the brain. Put another way, modern science provides compelling evidence that the complex biological system that is our body is near enough a deterministic machine as to make no practical difference. However, even if we set aside the issue of determinism, it seems the objection that plagued traditional compatibilism really plagues libertarianism as well. No sane libertarian wants to deny that your preferences, beliefs, and procedures for evaluating reasons for action are irrelevant to your choices, but regardless of whether you are your brain or you are your soul and regardless of whether determinism or indeterminism is true, a person will have preferences, beliefs, and procedures for evaluating reasons that in no way depend on the person. To think otherwise is to think that something can be the cause of itself, which is impossible.[17]

Perhaps the libertarian will reply that we don't need this kind of "self-causation" in order to explain how a person can be responsible for her choices. For example, Kane asks us to imagine a man who is on trial for murder, a man assumed to have a vicious character. Kane asserts that the man can be "ultimately" responsible for the murder only if he is ultimately responsible for his character. What is required for the man to be responsible for his character? Here is how Kane puts it:

> If there were factors or circumstances that entirely determined what he did, then to be ultimately responsible, he would have had to be responsible to some degree for one of those factors by virtue of earlier acts through which he formed his present character. ... But, of course, if *all* of our choices and actions were entirely determined by prior circumstances, we would have had to be responsible to some degree for some of these earlier circumstances by still earlier acts of ours, and so on indefinitely backwards in time—an impossibility for finite creatures like ourselves. At some point, if we are to be ultimately responsible for being what we are, there must be acts in our life histories in which parenting and society, genetic make-up, and other factors did not completely determine how we acted, but left something over for us to be responsible for then and there.[18]

Imagine that at some point in this man's past he wasn't yet vicious, and let's stipulate that his becoming vicious turned on a specific choice he made. Kane can grant that the man's beliefs, preferences, values, etc., were relevant to the choice he made, if only in the minimal sense that they inclined him to choose one way or another. Kane can also grant that those beliefs, preferences, values, etc., were the result of prior, determining factors (e.g., genetics and upbringing). But to be ultimately responsible for his character, it must be that although the man's beliefs, preferences, and values were relevant to his choice, they did not determine his choice.

But now the libertarian faces a dilemma. Suppose the balance of reasons favored the choice the man made (the choice that resulted in his having a vicious character).[19] If the balance of reasons did not determine his choice, then we face the lately discussed worry that libertarianism does violence to the relationship between choices and reasons for action that is characteristic of rational agency. Recall that my choice of salad was rational precisely because the balance of reasons favored it. In other words, given that the balance of reasons favored salad, the choice of soup would have made no sense. But now it looks like my being responsible for the choice of salad requires the availability of alternatives that I could not make sense of had they been actual. If my aim is to be rational, then it would seem that a situation in which irrationality is impossible is preferable to a situation in which irrationality is possible. Thus if the possibility of irrationality is a precondition for responsibility, then as a rational agent, I must judge that a world where responsibility does not exist is preferable to a world in which responsibility does exist.[20] This seems odd, at the very least.[21]

But now suppose that the balance of reasons didn't favor the choice the man made, not in the sense that the balance of reasons favored some other choice, but in the sense that the balance of reasons was compatible with more than one choice. As far as the balance of reasons goes, the man's choice is arbitrary; the choice might as well be settled by a coin flip. Kane recognizes the element of arbitrariness, and he correctly notes that to say a choice is arbitrary is not to say that one does not have a good reason for the choice. Rather, it is to say that one also has a good reason to make a *different* choice, and there is nothing one can point to that would show why one choice is better than the other. So whichever way I choose, I will have a good reason for choosing that way and am "willing to stand by and take responsibility for"[22] the choice I make. The problem with this reply is that it threatens to make indeterminism irrelevant to responsibility. The original worry was that if one does not have a reason to choose one way over another, then one ends up choosing for no reason, and choices made for no reason are not choices for which one is responsible. The flaw in this argument, of course, is that from the fact that one has no reason to choose A over B, it does not follow that one has no reason to choose A (and it does not follow that one has no reason to choose B). But if the point of the reply is to show how one can have a reason to choose as one did even if the choice was arbitrary, then it seems that what makes the person responsible for the choice is the fact that the choice was made for a reason. Libertarians like Kane will respond to this by saying that although choosing for reasons is a necessary condition for responsibility, it is not sufficient, and crucial to my being responsible for the choice I made is the fact that other choices were possible (in the sense of "possible" that conflicts with determinism). The presumed relevance of these alternatives is that without them there is nothing in the area of choice left to be settled, but if there is nothing in the area of choice left to be settled, then there is nothing in the area of choice left to be settled by *me*. Assume for the moment that it is reasonably clear what "the choice was settled by me" means. Why would the fact that a choice was settled by me imply that I am responsible for the choice? To be clear, this phrase does not merely mean "I made a choice," for the libertarian can grant that some choices we make are determined. So it must be that this phrase tells us something about the *conditions* in which a choice was made, or what was *involved* in the making of the choice, or how the choice *came about*. Moreover, it must be something about those conditions that accounts for how I can be responsible for the choice made. It is precisely here where the worry about arbitrariness resurfaces. Suppose I have a good reason to choose A, I have a good reason to choose B, and I choose A. This is consistent with the truth of determinism, so if determinism rules out the kind of responsibility we are after, I am not responsible for choosing A merely in virtue of having a good reason for choosing A. Responsibility requires that A and B are each possible (again, in the sense of "possible" that conflicts with determinism) and that the choice is settled by me. What was involved in settling it? Suppose you ask me "Why did you choose A instead of B?" I cannot give you a *good reason* for why I settled it the way I did, in the sense of "good reason" that is at work when I say that I have a good reason for choosing A. That is why the choice of A is arbitrary. It seems all I can say is "I just did." But it should now look utterly mysterious how "the choice was settled by me" can generate the kind of responsibility that is supposed to ground praise and blame, reward and punishment. This is not to say that I am not responsible for choosing A, of course. It's just that "the choice was settled by me" seems to contribute nothing to my responsibility beyond what is contributed by my choosing A *for a reason*.

I've said that while traditional compatibilists disagree with incompatibilists over how to analyze "could have chosen otherwise," each accepts that a person's choice is free only if the person could have chosen otherwise. I've also said each accepts that free will is a necessary condition for moral responsibility. This would imply that a person is morally responsible for a choice only if the person could have chosen otherwise. But is this true? In a widely influential paper, Harry Frankfurt argues that there are situations in which a person is morally responsible for what she chooses to do even though she could not have chosen otherwise than she did.[23] The kinds of cases Frankfurt uses to show this are cases in which a person makes a voluntary choice even though conditions are such that she would have made the choice had she begun to choose differently. However, because the conditions that make it true that the person could not have chosen otherwise play no actual role in bringing about the person's choice, the existence of those conditions seems irrelevant to whether the person is responsible. For example, suppose Jones wants Smith dead and chooses to kill him for that reason. Black also wants Smith dead, but Black is prepared to let Jones do the dirty work. To ensure that Jones kills Smith, Black inserts a chip into Jones' brain (unbeknownst to Jones) that will restore Jones' want to kill Smith if that want starts to waver. The want doesn't waver, however, so the chip is never activated. Jones wants to kill Smith and Jones' wanting to kill Smith explains Jones' choice to kill Smith. The presence of the chip guarantees that Jones could not have chosen otherwise than he did, but because the chip plays no part in the explanation of what happened, the fact that Jones could not have chosen otherwise is irrelevant. Given the irrelevance of the chip to what actually happened, the presence of the chip cannot bear on his responsibility. Thus Jones is responsible for killing Smith even though Jones could not have chosen to do otherwise than he did. Regardless of whether one ultimately agrees with Frankfurt, these cases force those who think "could have chosen otherwise" is relevant to moral responsibility—as I do—to explain *why* it is relevant.

At the beginning of the chapter I expressed my hope that a close examination of issues concerning the nature of mind, rationality, free will, and responsibility would make the seemingly obvious seem a lot less so. In fact, the challenge of reconciling free will and responsibility with determinism *or* indeterminism has proven so difficult that some philosophers have concluded that free will and responsibility are impossible,[24] or at the very least mysterious.[25] I believe this conclusion is premature, and in the next two chapters I will develop and defend an account that does justice to the way responsibility is connected to both free will and rationality. Along the way I will respond to the major challenges to traditional compatibilism identified in this chapter, including the challenge provided by Mr. Puppet, the seeming impossibility of deliberation in a deterministic world, and the claimed irrelevance to responsibility of "could have chosen otherwise"—both in general, and the traditional compatibilist reading in particular. Before doing this, however, we need to say more about the nature of mind and the kinds of confusion that talk of "the mind" can generate.

YOU ARE YOUR BRAIN

In Chapter 1 I said that fusing the manifest and scientific images requires rejecting mind-body dualism, and since the major aim of this book is to show how these images can be fused, it might be thought that the confusions talk of "the mind" generates are confusions that lead to dualism. To a certain extent this is true, but it will also turn out that there is plenty of room for confusion within the anti-dualist camp. Thus while fusion requires rejecting dualism, it also requires getting physicalism *right*. Let's start with the slogan signaling dualism's death: *you are your brain*. Read merely as a blunt rejection of dualism, most contemporary philosophers would accept it.[26] But there are a number of reasons why this slogan won't do as a statement of physicalism. For one, it is much too *narrow*. As a thesis about our mental lives, physicalism is the view that mental facts and properties depend solely on physical facts and properties. A word often used to express this dependence is *supervenience*. Supervenience can be characterized in terms of *necessary entailment*: Letting "P" stand for the set of all physical truths and "M" stand for the set of all mental truths, to say that the mental supervenes on the physical is to say that "If P, then M" is a necessary truth. The necessity at work in the supervenience relation can be captured in terms of possible worlds, and following David Braddon-Mitchell and Frank Jackson, we can say physicalism is the thesis that any world that is a physical duplicate of our world is a psychological duplicate of our world.[27] The basic idea is that the physical aspects of our world are *enough* to account for the mental aspects of our world. But notice that the statement of physicalism makes no reference to any *particular aspects* of the physical world. Here's why. Suppose facts about our mental life turn out to depend solely on physical facts about the Crab Nebula. In that case, while it would be false that you are your brain, it would still be true that mental facts depend solely on physical facts and so physicalism would still be true. We think the Crab Nebula scenario is ridiculous, of course, but we want to characterize physicalism in a manner that includes all the ways in which mental facts might depend on solely physical facts, and the way to do that is by stating it in terms of all the physical facts.[28]

Another reason why the slogan won't do as a statement of physicalism is that it may tempt us to treat minds as *things*. The idea that minds are things largely derives from Descartes. According to Descartes, the mind is "the thing that thinks." Descartes went on to argue that no physical thing could think, so the thing that thinks must be a nonphysical thing. Cartesianism is thus the view that (i) minds are things, and (ii) minds are nonphysical. A good deal of criticism has been aimed at (ii). To see why, note first that though Descartes believed that mind and body are different kinds of things, he also believed that they causally interact. For example, my intention to make dinner (mental) causes my body to move toward the kitchen (physical), and the stubbing of my toe (physical) causes me to feel pain (mental). Unfortunately, Descartes had a conception of causality according to which causality requires contact and contact requires extension. If the mind literally lacks extension—for Descartes, extension is a property of bodies, not minds—then the mind cannot come into contact with a body and so cannot causally interact with a body. Mind-body interaction is so obvious that we should reject any hypothesis about the nature of mind that cannot account for it. To many, Descartes' dualism cannot account for it and thus should be rejected.

Greene and Eagleman think that advances in neuroscience will make it virtually impossible to deny that all thought and behavior is rooted in biology. While the neural dependence of thought does not show that dualism is metaphysically impossible, it is plausible to suppose that the best explanation of the neural dependence of thought is that thought *is* a neural phenomenon. Why do my thoughts alter when my brain alters? Because *you are your brain*. Note then that Greene and Eagleman can agree with Descartes that the "I" is a thinking thing—a mind—even if they claim that you are your brain. You are a thinking thing, but it turns out that the thing that thinks is a brain. If this is what we are being invited to accept, then we can say that while (i) is true (minds are indeed things), (ii) is false (the mind is physical, it turns out).

Daniel Dennett calls the view according to which minds are physical things "Cartesian materialism."[29] The label is pejorative, but not because the position is a version of physicalism. The problem is with the idea that minds are things. The rejection of the view that minds are things owes much to the work of Gilbert Ryle. According to Ryle, while "I have a mind" and "I've lost my mind" appear grammatically similar to statements like "I have a wallet" and

"I've lost my wallet," the grammar is misleading. When we say that a creature "has a mind" we are attributing to the creature certain kinds of *capacities, dispositions, or abilities*.[30] To say that a person has a mind is like saying that a car has a certain horsepower. Horsepower is not a thing in the way the parts of a car are things; rather, talk of horsepower is talk of how much work the car can do over time. We cannot see horsepower in the way that we can see pistons and spark plugs, but this does not mean that horsepower is some kind of spooky, nonphysical stuff. It isn't any kind of stuff. Of course, even though horsepower is not among the parts of a car, a car's horsepower depends on the parts and how the parts are arranged. In this way we can still pose a meaningful issue about *whether horsepower is physical*. To say horsepower is physical is to say that facts about horsepower depend solely on physical facts, e.g., physical facts about car parts and their arrangement. Similarly, to say that minds are physical is to say that the capacities picked out by our talk of "the mind" depend solely on physical parts and their arrangement.

Let's call the view according to which having a mind amounts to possessing certain kinds of capacities *Ryleanism*. If having those capacities depends solely on physical facts, then Ryleanism appears to provide a physicalist friendly alternative to Cartesian materialism. Most contemporary philosophers of mind endorse neither view, however, instead embracing some version or other of what is called *functionalism*. There are two basic, albeit related, routes to functionalism. The first route is through the analysis of mental concepts. Historically, functionalists of this stripe were largely reacting to philosophical behaviorism. Philosophical behaviorists were also interested in the analysis of mental concepts, and the core tenet of philosophical behaviorism was that mental concepts are dispositional concepts. An example often used to illustrate the idea is *being fragile*. To say that an object is fragile is to say something about how the object would behave in certain conditions, e.g., it would break if lightly struck. Mental concepts are like that, said the philosophical behaviorist, e.g., in saying that a person is in pain, we are saying something about how the person would behave in certain conditions. Pain is not the same as behavior, of course, just as being fragile is not the same as breaking.[31] Rather, it is a *disposition* to behave. Functionalists interested in the analysis of mental concepts agreed with philosophical behaviorists that there is a conceptual connection between mental states and behavior, but they disagreed about the nature of this connection. According to "causal role" functionalism,[32] concepts of mental states are not dispositional concepts, but rather causal concepts, e.g., the concept of pain is the concept of an inner state that is apt to cause screaming and writhing. On this view, each mental concept (the concept of belief, the concept of desire, and so forth) is a concept of an inner state that satisfies a particular causal profile (where these profiles include causal relations to other inner states and behavior). Mental states are differentiated by the causal profiles associated with each mental state concept, and on this view, to say that a creature has a mind is to say that the creature has states that satisfy the causal profiles associated with various mental concepts.

But what is the nature of these inner states, the states that allegedly satisfy the causal profiles associated with our mental concepts? Functionalists had a neat answer to this question. As far as the analysis of mental concepts goes, these inner states could be anything—states of the brain, states of a nonphysical soul, states of a silicon robot, etc. In this way, functionalism was officially neutral on the issue of physicalism. In practice, however, functionalists were betting that the states satisfying these causal profiles would turn out to be brain states, and if they did, this would allow us to *identify* mental states with brain states.[33] Functionalism provided a two-step strategy for deriving such identities. The first step is the conceptual analysis of mental concepts: mental concepts are concepts of states that play a certain functional role (satisfy a particular causal profile). The second step is the empirical discovery that the states satisfying these causal profiles are states of the brain. If neuroscience reveals that c-fiber stimulation satisfies the causal profile for pain, then we should say that pain *is* c-fiber stimulation.

Because the version of functionalism I've just described is primarily a thesis about the analysis of mental concepts, it is often called "analytic functionalism." There is a second route to functionalism, however, one that differs from analytic functionalism in important ways. Historically, functionalists of this second kind were not primarily interested in the analysis of mental concepts but rather the implications of formal logic and computing theory for the science of the mind. Inspired by the seminal work of Alan Turing (and others), many philosophers began to take seriously the hypothesis that the mind is a computer. Artificial intelligence was founded on this hypothesis, and one of the major commitments of this idea is that to understand what makes a cognitive system cognitive, we need a "level of analysis, which is independent of physics, yet is mechanistic in spirit. It makes possible a science of structure and function divorced from material substance, while at the same time it avoids the retreat to behaviouristic peripheralism."[34] A science of the mind divorced from material substance does not signal a return to dualism, however, for the claim here isn't that computation takes place in some nonphysical stuff. Rather, the claim is that to describe a system as a computing system is to describe the system at a level of abstraction that ignores many of the physical details of the system. In this way, systems that differ physically may be the same computationally, and so if cognition is computation, then a science of cognition can safely ignore many of the physical details of this or that system. In particular, if cognition can take place in nonbiological machines, e.g., my laptop, then the understanding of cognition we are after *won't* be provided by neuroscience.

Reflection on this second route to functionalism suggests yet a third reason why *you are your brain* won't do as a statement of physicalism. If we read this as the claim that mental properties are identical to neural properties, then this claim says something stronger than what physicalism says. Cognitive science takes seriously the possibility that nonbiological machines can have a mental life, i.e., it takes seriously that mental phenomena are multiply realizable. The idea of multiple realization is fairly straightforward. Consider a doorstop. You can make a doorstop out of rubber, but you can also make a doorstop out of wood. So the property of being a doorstop cannot be identified with the property of being a piece of wood configured thus and so. This is consistent with each *instance* of the property of being a doorstop being identical to an *instance* of some physical property or other, but from the fact that each doorstop is an instance of some physical property or other, it doesn't follow that there is a single physical property of which each door stop is an instance. As far as physicalism about doorstops goes, all we need is that each instance of being a doorstop is identical to an instance of some physical property or other. The same applies to the mind. If the project of "reducing" mind to body is understood as the project of discovering *property* identities, then while the success of such a project would be sufficient to vindicate physicalism, it isn't necessary to vindicate physicalism. In short, supervenience is weaker than reduction, and physicalism requires only supervenience.

Or so some functionalists claim. I will have a lot more to say about each version of functionalism, but for now I want to highlight some difficulties facing functionalism, difficulties we will have to confront, I argue, if we are to fuse the manifest and scientific images. While I have identified two routes to functionalism, philosophical discussions of functionalism often conflate these two routes. This conflation has led to confusion. Consider again the strategy for deriving identity claims provided by analytic functionalism, the one that led us to conclude that pain is c-fiber stimulation (where this is understood as a claim of property identity). The claim that pain is c-fiber stimulation is not analytic, for it is not part of the concept of pain that pain is c-fiber stimulation. The concept of pain is the concept of an inner state that satisfies a particular causal profile, but this tells us nothing about the state itself (except that it satisfies the causal profile). So we have the concept of pain, on the one hand, and what pain is, on the other. But as Saul Kripke taught us,[35] if 'pain = c-fiber stimulation' is true, it is *necessarily* true, and if it is necessarily true, then it is impossible for a creature that lacks c-fibers to be in pain. This applies, mutatis mutandis, to all mental states, e.g., if the belief that Paris is the capital of France is identical to z-fiber stimulation, then it is impossible for a creature that lacks z-fibers to have this belief. The second route to functionalism, however, envisions a science of the mind divorced from material substance, i.e., this science wants to allow that physically disparate systems can have the *same* mental life. The functionalism that leads to identifying mental properties with physical properties seems inconsistent with the functionalism that leads to a science of the mind divorced from material substance.

There are a number of ways one might try to overcome this tension. One way is to relativize or restrict the relevant property identity claims.[36] For example, suppose we come across a race of Martians and discover that each member of this race has an inner state satisfying the causal profile of pain. Call this state MP1. We might then say pain-in-humans is c-fiber stimulation, while pain-in-Martians is MP1. But this tactic has its costs, for it will turn out that there is not a single type of state (property) that both humans and Martians can be in; after all, c-fiber stimulation *is not* MP1. If we further assume that scientific categories correspond to state types (properties), then instead of having a single science of mind, we will have a science of the human mind, a science of the Martian mind, a science of the Macintosh mind, etc. But then we don't really have a science of the mind that is *divorced* from material substance.

But hope is not yet lost, for in fact there is a state type here that Martians and humans share: the state of having a state that satisfies the causal profile of pain. On this view pain is not c-fiber stimulation, *even in humans*. Instead, pain—the state type—is the state of having a state that satisfies a certain causal profile. Humans have a state that satisfies this causal profile, and that state is c-fiber stimulation. Martians have a state that satisfies this causal profile, and that state is MP1. Pain is thus a "second-order" or "higher-order" state, one characterized in terms of "first-order" or "lower-order" states. The second-order state is realized differently in humans and Martians, but pain isn't the state that realizes a causal profile. Rather, it is the state of having a state that realizes a causal profile. The view according to which mental states are second-order states is often called "role functionalism," whereas the view according to which mental states are first-order states is often called "realizer functionalism."[37]

It is easy to lose the point of these dialectical maneuvers, and perhaps you are hoping that I will say that role functionalism is great and everything is OK. I will be saying this—or something *close* to this—but role functionalism must first clear a big hurdle. I want to get at this hurdle by placing the debates over functionalism in the larger context of what fusion requires. In "Empiricism and the Philosophy of Mind," Sellars notes that "in characterizing an episode or a state as that of *knowing*, we are not giving an empirical description of that episode or state; we are placing it in the logical space of reasons, of justifying and being able to justify what one says."[38] When we locate an episode in the space of reasons we provide a kind of explanation of that episode, and the crucial feature of such "reasons explanations" is the way they justify, or make sense of, or make intelligible, the episodes being explained. While Sellars talks about knowledge in this passage, it is clear that actions are also placed in the space of reasons. For example, suppose I ask "Why did you stop at the red light?" and you answer "Because my mother's favorite color is blue." I would reject this explanation, but not because empirical investigation failed to confirm a regularity linking familial color preferences and driving behavior (even if I assume empirical investigation would fail to confirm this). Rather, I would reject this explanation on the grounds that the fact that your mother's favorite color is blue does not justify, or make sense of, or make intelligible, your car stopping behavior.

If characterizing an episode as a state of knowledge (or a type of action) is a matter of locating that episode in the space of reasons, into what other kinds of "spaces" might we locate episodes, i.e., into what kind of "space" do we locate an episode when we give an *empirical description* of the episode? If the space of reasons has its home in the manifest image, then it is plausible to suppose that the space we are looking for has its home in the scientific image. Following John McDowell, we can say that the conception of nature provided by the scientific image is one "according to which something's way of being natural is its position in the realm of law,"[39] or what Peter Menzies calls "the space of causes."[40] The possibility of fusing the manifest and scientific images of human beings would thus seem to require the possibility of locating individual episodes in both the space of reasons *and* the space of causes.

There are a number of considerations that cast doubt on whether any episode can be so located, however.[41] For my purposes, I will be focusing on just one way of developing this doubt, one that raises troubles for the explanation of action. The exposition here will be brief, with further elaboration coming in Chapter 6. For now, I merely want to convey a sense of the problem. Consider the connection between pain and pain behavior. We say pain *causes* screaming and writhing, and we explain the screaming and writhing by citing its cause. Lots of explanation in science works like this, but we also know that science requires empirical investigation. In their search for causes, scientists actually have to observe the world. But now consider role functionalism about pain. According to role functionalism, being in pain is being in a state that causes (among other things) screaming and writhing. But then "pain causes screaming and writhing" is equivalent to "being in a state that causes screaming and writhing causes screaming and writhing." Unlike the causal explanations we expect to find in science, this explanation is trivial and uninformative.[42] If that's right, then it looks like every explanation that role functionalism generates will be equally trivial and uninformative. Because I will be defending a version of role functionalism and the idea that we can locate an action in both the space of reasons and the space of causes, the burden is on me to show what is wrong with this argument.

NOTES

1. Plato, *Euthyphro*, trans. G.M.A Grube, in *Plato: Five Dialogues*, ed. John Cooper (Indianapolis: Hackett, 2002), 1–20.

2. Jeffrey Rosen, "The Brain on the Stand," *New York Times Magazine*, March 11, 2007, accessed June 2, 2015, http://www.nytimes.com/2007/03/11/magazine/11Neurolaw.t.html.

3. Rosen, "The Brain on the Stand."

4. Greene and Cohen, "Neuroscience Changes Nothing and Everything," 1780.

5. Greene and Cohen, "Neuroscience Changes Nothing and Everything," 1780.

6. David Eagleman, "The Brain on Trial," *Atlantic*, July/August 2011, accessed May 3, 2015, http://www.theatlantic.com/magazine/archive/2011/07/the-brain-on-trial/308520/.

7. Eagleman, "The Brain on Trial."

8. Eagleman, "The Brain on Trial."

9. Eagleman, "The Brain on Trial."

10. Eagleman, "The Brain on Trial."

11. Richard Nisbett and Timothy Wilson, "Telling More Than We Can Know: Verbal Reports on Mental Processes," *Psychological Review* 84 (3) (1977): 231–259.

12. Richard Taylor, "Libertarianism: Defense of Free Will," in *Introduction to Philosophy: Classical and Contemporary Readings*, ed. James Fieser and Louis Pojman (New York: Oxford University Press, 2008), 405.

13. Traditional compatibilism traces back to at least Hume (*A Treatise of Human Nature*, ed. P.H. Nidditch. Oxford: Clarendon Press, 1978), and its 20th century defenders include A.J. Ayer (Philosophical Essays. New York: St. Martin's Press, 1954.) and Walter Stace (*Religion and the Modern Mind*. London: MacMillan & Co., 1953), among others.

14. Immanuel Kant, *The Critique of Practical Reason*, in *Great Books of the Western World* 42 (Chicago: University of Chicago, 1952), 332.

15. This argument can be viewed as an instance of what Peter van Inwagen calls "the Principle": "Suppose it's an untouchable fact that *p*. And suppose also that the following conditional (if-then) statement expresses an untouchable fact: if *p*, then *q*. It follows from these two suppositions that it's an untouchable fact that *q*" ("The Powers of Rational Beings: Freedom of the Will," in *Philosophy: The Quest for Truth*, ed. Louis Pojman and Lewis Vaughn Metaphysics (New York: Oxford University Press, 2014), 420).

16. Robert Kane, "Free Will: Ancient Disputes, New Themes," in *Reason and Responsibility*, ed. Joel Feinberg and Russ Shafer-Landau (Belmont, CA: Wadsworth, 2002).

17. Galen Strawson, "The Impossibility of Moral Responsibility," *Philosophical Studies* 75 (1–2) (1994): 5–24.

18. Kane, "Free Will," 502.

19. It may seem odd to imagine that the balance of reasons could favor this choice, given that it resulted in the man becoming vicious. I'll have more to say about this in Chapter 4, but for now we can grant the oddity and simply note that it is not relevant to the argument.

20. On the condition that in the world in which responsibility does not exist the explanation of its nonexistence is that irrationality is not possible in that world (in the sense of possibility that the incompatibilist has in mind).

21. As I will show in Chapter 4, this is not odd if the world in which no one is responsible is also a world in which everyone is perfectly rational. It will turn out that responsibility has relevance only in worlds in which people are not perfectly rational, and when we appreciate why this is so, we will also appreciate that the traditional compatibilist's account of free will makes sense of this kind of responsibility.

22. Kane, "Free Will," 512.

23. Harry Frankfurt, "Alternate Possibilities and Moral Responsibility," *Journal of Philosophy* 66 (23) (1969): 829–839.

24. Strawson, "The Impossibility of Moral Responsibility."

25. van Inwagen, "The Powers of Rational Beings."

26. To the extent that dualism is still considered a viable option, it is mostly a position on the nature of phenomenal consciousness. While this position has a number of able defenders—most notably, David Chalmers—it is not the majority view. More importantly, most contemporary arguments for dualism do not turn on those features of our mental lives that are most clearly relevant to law. For example, Chalmers divides problems of consciousness into *easy* and *hard* (David Chalmers, "Facing Up to the Problem of Consciousness," *Journal of Consciousness Studies* 2 (3) (1995): 200–219). The easy problems concern such things as the ability to integrate information, to report mental states, to focus attention, to deliberately control behavior; by contrast, the hard problem concerns phenomenal consciousness, or the "what it is like" aspect of various perceptual states. Arguably, all the features of our mental life that are relevant to law pose at best easy problems, and Chalmers does not think the easy problems pose a challenge to physicalism.

27. David Braddon-Mitchell and Frank Jackson, *Philosophy of Mind and Cognition* (Malden, MA: Blackwell, 2008), 29.

28. Again, in including all the physical facts we are not assuming that all the physical facts actually matter to the mental facts. The point of including all the physical facts is that it guarantees that the relevant physical facts are included, even if we don't know which of the physical facts are relevant.

29. Daniel Dennett, *Consciousness Explained* (Boston: Little, Brown and Co., 1991).

30. Gilbert Ryle, *The Concept of Mind* (London: Hutchinson, 1949).

31. And while we typically think of fragility as a somewhat stable disposition of an object, this is not a strict requirement, i.e., the dispositions at issue here can be short-lived (as pains sometimes are, thankfully).

32. David Armstrong, *A Materialist Theory of the Mind* (London: Routledge, 1968).

33. Armstrong, *A Materialist Theory of the Mind*; David Lewis, "An Argument for the Identity Theory," *Journal of Philosophy* 63 (1966): 17–25.

34. Zenon Pylyshyn, "Complexity and the Artificial Study of Human Intelligence," in *Philosophical Perspectives in Artificial Intelligence* (Atlantic Highlands, NJ: Humanities Press, 1979), 24–25. If it is fair to say that analytic functionalism was a reaction to the perceived shortcomings of philosophical behaviorism, then it is

equally fair to say that functionalism in psychology was a reaction to the perceived shortcoming of *scientific* behaviorism, e.g., the kind of behaviorism associated with B.F. Skinner.

35. Saul Kripke, *Naming and Necessity* (Oxford: Blackwell, 1980).

36. David Lewis, "Mad Pain and Martian Pain," in *Readings in the Philosophy of Psychology*, ed. Ned Block (Cambridge, MA: Harvard University Press, 1980), 216–222.

37. Brian McLaughlin, "Is Role-Functionalism Committed to Epiphenomenalism?" *Journal of Consciousness Studies* 13 (1–2) (2006): 39–66.

38. Wilfrid Sellars, "Empiricism and the Philosophy of Mind," in *Science, Perception, and Reality* (Atascadero, CA: Ridgeview Publishing Company, 1963).

39. John McDowell, *Mind and World* (Cambridge, MA: Harvard University Press, 1994), 74.

40. Peter Menzies, "Reasons and Causes Revisited," in *Naturalism and Normativity*, ed. Mario de Caro and David Macarthur (New York: Columbia University Press, 2010), Kindle edition, 406.

41. Consider the familiar case of forming a belief as a result of some perceptual experience, e.g., forming the belief that there is a cat in the room as a result of hearing a meow. Oftentimes such beliefs seem justified, and when they are, it is intuitive to think that the belief is justified by the experience, e.g., hearing the meow. What must the relationship between the experience and the belief be in order for the experience to justify the belief? Suppose we try to locate this episode in the space of causes. The story might go something like this: kitty cat emits sound waves (it meows). The sound waves impact my auditory system, causing me to have a perceptual experience (I hear the meow). In turn, this experience causes me to come to believe that there is a cat in the room. The problem with this story, however, is that it appears to provide "exculpations where we wanted justifications" (McDowell, *Mind and World*, 8). McDowell's worry about this purely causal story is that, given the current local environment and the way I am built, I cannot help but form the belief that there is a cat in the room, but while this may show that I cannot be *blamed* for believing as a do, it does not show that I am entitled to the belief, i.e., it does not show that the belief is *warranted*. For a critical reply, see Jerry Fodor, "Review of John McDowell's *Mind and World*," in *In Critical Condition* (Cambridge, MA: MIT Press, 1998), 3–8.

42. Robert Rupert, "Functionalism, Mental Causation, and the Problem of Metaphysically Necessary Effects," *Nous* 40 (2) (2006): 256–283.

Chapter 3

Spiderman, Doing Whatever a Spider Can

The perhaps most famous depiction of responsibility in recent popular culture is provided by the August, 1962 issue of *Amazing Fantasy*, the issue that introduced Spiderman to the world (Figure 3.1). The line linking power and responsibility has become a familiar slogan, and in the *Spiderman* movie franchise the relationship is expressed as "with great power comes great responsibility." This formulation is wonderfully ambiguous—there are at least three plausible ways to interpret it—but as I will argue, what makes the ambiguity wonderful is that each interpretation captures an important aspect of the concept of responsibility. The goal of this chapter is to identify those aspects and reveal how they are related. Once we do, we will see that the issue of determinism is largely irrelevant to our ordinary understanding of responsibility.

THREE SENSES OF RESPONSIBILITY

P-responsibility

Sometimes the word 'responsible' is used as a synonym for 'cause,' as when the National Hurricane Center states that the New England Hurricane of 1938 "was responsible for 600 deaths and $308 million in damage in the United States."[1] However, insofar as the concept of responsibility we are interested in is intimately linked to praise and censure, reward and punishment, the above sense of 'responsible' isn't what we are after. The reason isn't just that phenomena like hurricanes are not suitable targets for praise and censure, reward and punishment. Humans, too, can cause things to happen without being responsible in the sense we are after. Consider the insanity plea. If a

Figure 3.1 A cartoon panel from *Amazing Fantasy*. (no.15); August, 1962. Sincere thanks to Marvel Entertainment, LLC for permission to use this work.

person being tried for first degree murder successfully pleads not guilty by reason of insanity, the law views this person as not legally responsible for the killing even if the person was the killer, i.e., even if the person was the cause of the victim's death. Evidently the kind of responsibility we are after has something to do with a sort of agency and capacity for intentional action that is not possessed by hurricanes or the legally insane. For example, Michael Levin analyzes 'Agent A is responsible for action X' as 'X happened because A wanted X to happen.'[2] However, it seems that there are cases in which an Agent A is responsible for an action X even when X didn't happen because A wanted X to happen. Spiderman judges himself at fault for Ben's death and accepts responsibility for it. If action X here is the murder of Ben, then we can say that Spiderman accepts responsibility for action X.[3] But Spiderman didn't want Ben to be murdered, so Ben's murder didn't happen because Spiderman wanted it to happen. Thus, insofar as we agree with Spiderman's acceptance of responsibility, this provides a clear counterexample to Levin's analysis. The law itself furnishes us with other counterexamples. For instance, parents can be held civilly liable for what their children do. If my 11-year old son sprays graffiti on the walls of his school, a court can order me to pay for the cleanup. In this way I am responsible for the spraying of graffiti even though I didn't want the spraying of graffiti (and thus the spraying of graffiti didn't happen because I wanted it).

These counterexamples suggest an amendment to Levin's analysis, an amendment that captures the fact that whether the action occurred had something to do with whether an agent exercised certain of the agent's powers. Spiderman had the power to prevent the murder, so he is responsible for the murder, and because parents have the power to prevent their children from performing certain actions, e.g., by restricting their access to means and opportunities, they can be responsible for what their children do. Regardless of how we amend the analysis, however, Levin is surely correct that there is a sense of 'responsible' that concerns the causal powers of agents. Let's call the senses of 'responsible' and 'responsibility' that concern such powers *P-responsible* and *P-responsibility*.

On one reading of "with great power comes great responsibility," Spiderman's realization is that responsibility is a direct function of power, i.e., that responsibility is P-responsibility. The left side of the panel certainly suggests this: Spiderman is at fault—is responsible—for Ben's death *because* he could have prevented the death—it was within his power to prevent it. But a closer look at the right side of the panel suggests that this doesn't tell the whole story. We are told Spiderman becomes aware "that in this world, with great power there must also come—great responsibility!" The qualification "in this world" and the phrasing "there must also come" suggest that there can be circumstances in which one has great power but does *not* have great responsibility. If that's right, then the phrase cannot be claiming that responsibility is a function of power, for on that reading it would be contradictory to say that Spiderman had the power to prevent the death but was not responsible for the death. Moreover, if great power were sufficient for great responsibility, then the point could be more clearly put as "with great power there is great responsibility!" In the context of the claim that we are witnessing the birth of a legend, "with great power comes great responsibility" communicates that there are facts about our world that enjoin Spiderman to become *a certain kind of person*.

D-responsibility

There are ordinary uses of 'responsible' that do not characterize an agent's relation to a specific action that occurred. When we praise a teenager for being a "responsible young woman," for example, we appear to be attributing to this teenager a character trait, not a relation to something that happened. Following Aristotle, we can think of character traits as dispositions to behave in certain kinds of ways, e.g., the courageous person is disposed to act bravely. Let's call the dispositional sense of 'responsible' and 'responsibility' *D-responsible* and *D-responsibility*.[4]

The sense of responsibility at play in D-responsibility is perhaps most vividly expressed by Friedrich Nietzsche. In the second essay of *On the*

Genealogy of Morals, Nietzsche states that the real "problem" of human beings is "to breed an animal *with the right to make promises*."[5] The person who has the right to make promises is constituted "so that between the original 'I will,' 'I will do,' and the actual discharge of the will, its *act*, a world of strange new things, circumstances, even acts of will may be interposed without breaking this long chain of will."[6] To breed such an animal, "Man himself must first of all have become *calculable, regular, necessary*, even in his own image of himself, if he is to be able to stand security for *his own future*, which is what one who promises does!"[7] Nietzsche thinks that the development of an animal that is entitled to make promises is also "the long story of how *responsibility* originated."[8] The end result of this development is "the sovereign individual" who possesses "a consciousness of his own power and freedom" and for whom "the proud awareness of the extraordinary privilege of *responsibility*, the consciousness of this rare freedom, this power over oneself and over fate, has in his case penetrated to the profoundest depths and become instinct."[9]

Those who have the right to make promises are responsible in the sense that such people are *reliable*; they are people who can be *counted on*. The birth of a legend is the birth of a hero, and with this comes the suggestion that we can count on Spiderman to "do whatever a spider can," i.e., we can trust that he will *exercise* his powers so as to prevent future crimes. Responsibility is not merely about the bare possession of powers, but how and when one exercises those powers.

While rendering responsibility as D-responsibility certainly brings us closer to a satisfactory understanding of the pivotal phrase "with great power comes great responsibility," something is still missing. The panel does not convey simply that our world requires legends; it tells us that we are witnessing the *birth* of one. Might there be yet a third sense of responsibility at work here, one that would explain how D-responsibility can be born from P-responsibility?

A-responsibility

Someone can be P-responsible for an action even if they are not a D-responsible person. For example, consider a student who routinely fails to do her homework. Such a person is an irresponsible student, i.e., is not D-responsible, qua student. But insofar as it is within the student's power to turn in her homework, she is P-responsible for not turning in her homework. P-responsibility and D-responsibility are importantly related, however. A good illustration of this is provided by Aristotle's account of how someone becomes virtuous. According to Aristotle, virtues are dispositions, and one acquires virtue by doing virtuous things. Consider again what is involved in being a responsible student. A responsible student is a student who reliably does what a student ought to do, and what a student ought to do is what a student needs to do to be a good student. Moreover, the fact that the responsible student reliably does what a student ought to do is explained, at least in part, by the fact that the responsible student knows what a student ought to do. Without such knowledge it would be a vast coincidence that the student reliably does what a student ought to do. By contrast, an irresponsible student is a student who does not reliably do what a student ought to do. However, the irresponsible student is not typically characterized as lacking knowledge of what a student ought to do. Moreover, the irresponsible student is not typically characterized as lacking the power to do what a student ought to do. Indeed, if the irresponsible student lacked the power to do what a student ought to do, the irresponsible student couldn't become a responsible student by doing what a student ought to do. So why does the responsible student, but not the irresponsible student, reliably do what a student ought to do? Aristotle recognized the critical role that pain and pleasure play in the possession of virtue, and put simply, it pleases the responsible student to do what a student ought to do but it doesn't please—and might even pain—the irresponsible student to do what a student ought to do. Aristotle's insight, however, was that the presence (or absence) of the pleasure (or pain) accompanying our actions can change as a function of how often we perform certain action types. The more the irresponsible student does what a student ought to do, the *less painful* or *more pleasurable* it is for the student to do what a student ought to do, and the *less painful* or *more pleasurable* it is for the student to do what a student ought to do, the *more likely* it is that the student will do what a student ought to do. The net effect is the inculcation of the very disposition that characterizes the responsible student.

Returning to Spiderman, our challenge is to explain how D-responsibility can emerge from P-responsibility, and as the discussion in the previous paragraph suggests, this development requires that P-responsible agents be responsible in yet a third sense, a sense in which 'responsible' means 'is able to respond' and 'is able to respond' means 'is able to adapt.' Let's call the adaptive sense of 'responsible' and 'responsibility' *A-responsible* and *A-responsibility*.

Of course, being able to adapt is not the same thing as actually adapting, so we still need to explain why Spiderman in fact adapts, i.e., we need to identify the factors that account for the birth of a legend. Chief among these is the fact that Spiderman *takes* responsibility for Ben's death. Taking responsibility involves judging oneself P-responsible, and the relevance of this judgment is made clear when we consider how Spiderman would have reacted if he had not judged himself P-responsible for Ben's death. Given that he cared deeply about Ben, we can imagine that he would be grief-stricken. But when we take a close look at Spiderman's face, we don't see a person who is merely grief-stricken; we see someone who is racked with guilt, remorse, and perhaps shame. These reactions would be appropriate only if Spiderman viewed Ben's death as resulting from a failure on his part, i.e., only if Spiderman judged himself P-responsible for the death. The judgment of P-responsibility is not sufficient for the kind of adaptation we're after, however. To see this, consider how Spiderman would have reacted if he hadn't cared about Ben or didn't otherwise view the death negatively. More generally, consider how Spiderman would have reacted if he were indifferent to the safety and liberty of ordinary citizens. In such a case Spiderman might judge himself P-responsible for the death, but he would not view his "failure" to prevent Ben's death as a "failure" that warrants remorse or shame. If either of these conditions obtained—Spiderman did not judge himself P-responsible for the death or was indifferent to the death—we would not believe that we were witnessing the birth of a legend. Consider the scenario in which Spiderman is indifferent. If he is indifferent, then it is nothing to him whether villains threaten the safety or liberty of ordinary citizen, and so we would have no reason to think that he will exercise his power to stop these villains. But even if Spiderman cares deeply about the safety and liberty of ordinary citizens, there is no reason to think he will exercise his power to stop villains if he does not think himself P-responsible for what happens to these citizens. *Ben's death had nothing to do with me*, Spiderman might think, and if he does think this, then he may similarly think that whether ordinary citizens are harmed has nothing to do with him. In that case, there would be no reason to think he will act differently if faced with a similar situation in the future; *had nothing to do with me* means *didn't depend on how I acted*, so there would be nothing to prompt changes in Spiderman's future behavior.[10] We can thus see that Spiderman's guilt, remorse, and shame are the result of caring deeply about Ben *and* judging himself P-responsible for Ben's death, and it is the *nature of his reaction* that lends credence to the claim that a legend is born. His reaction can lend such credence because his reaction affects how he will think and feel in future situations, and it does so in a way that makes it more likely that he will exercise his power in those situations. A-responsibility explains how D-responsibility emerges from P-responsibility.

But what does this have to do with *moral* responsibility, the kind of responsibility that explains how people can *deserve* praise and blame, reward and punishment? I claim that the three kinds of responsibility introduced above provide us with the resources we need to make sense of moral responsibility. The task of the next two sections is to make good on this claim.

MANIFEST RESPONSIBILITY

At the outset I said that the manifest image conceives of human beings as free, rational, responsible agents who are subject to moral and epistemic norms, norms that legitimize praise or blame, reward or punishment. It is now time to take a closer look at the details of this image. Wilfrid Sellars coined the phrase "manifest image," and Sellars himself wrote that the manifest image "corresponds to the world as conceived by P. F. Strawson … a framework in which the distinctive features of persons are conceptually irreducible to features of nonpersons, e.g., animals and merely material things."[11] One distinctive feature of persons is their participation in interpersonal relationships, and in his seminal paper "Freedom and Resentment," Strawson emphasizes the role that *reactive attitudes* play in such participation.[12] According to Strawson, the reactive attitudes—including contempt, censure, blame, praise, gratitude, and a host of others—"rest on, and reflect, an expectation of, and demand for, the manifestation of a certain degree of goodwill or regard on the part of other human beings towards ourselves"[13] and "are essentially natural human reactions to the good or ill will or indifference of others towards us, as displayed in *their* attitudes and actions."[14] Morality is connected with the "generalized aspect of the demand,"[15] as is reflected in our reactions of indignation and disapproval when demands for goodwill are not expressed toward others:

> One who experiences the vicarious analogue of resentment is said to be indignant or disapproving, or morally indignant or disapproving. What we have here is, as it were, resentment on behalf of another, where one's own interest and dignity are not involved; and it is this impersonal or vicarious character of the attitude, added to its others, which entitle it to the qualification "moral."[16]

In addition to the attitudes directed toward others, there are also what Strawson calls self-reactive attitudes, including "feeling bound or obliged (the 'sense of obligation'); feeling compunction; feeling guilty or remorseful or at least responsible; and the more complicated phenomenon of shame."[17] Strawson also emphasizes the nonincidental way that these attitudes are related to thought and behavior. For example, indignation and disapprobation "tend to inhibit or at least to limit our goodwill towards the object of these attitudes, tend to promote an at least partial and temporary withdrawal of goodwill,"[18] and

> Just as the other-reactive attitudes are associated with a readiness to acquiesce in the infliction of suffering on an offender, with the "institution" of punishment, so the self-reactive attitudes are associated with a readiness on the part of the offender to acquiesce in such infliction *without* developing the reactions (e.g. of resentment) which he would normally develop in the infliction of injury upon him: i.e. with a readiness, as we say, to accept punishment as "his due" or as "just."[19]

Strawson's lengthy discussion of the reactive attitudes and their effects on thought and behavior is not intended to tell us something we do not already know; rather, "The object of these commonplaces is to try to keep before our minds ... what it is actually like to be involved in ordinary inter-personal relationships."[20] For Strawson, the point of these reminders is to draw our attention to the fact that it is within the context of our actual moral practice that we come to understand concepts like desert and responsibility. When we pay close attention to these contexts, the accounts we develop of these concepts appear somewhat anodyne. For example, consider what is involved in saying that someone deserves our moral disapproval. To say that someone deserves our disapproval is just to say that the person is worthy of it, or the person merits it. To disapprove of something is to view it as something one ought not to do. If I demand that you express goodwill toward me and if that demand is legitimate, then you ought to express goodwill toward me. If you don't express goodwill toward me, then you are doing something you ought not to do. My disapproval is therefore justified in that you are doing something you ought not to do and I am viewing it as something you ought not to do. I see it that way, and it is that way. That is what makes it appropriate, i.e., you deserve my disapproval. If this "explanation" does not seem particularly deep or illuminating, if it sounds like an unnecessarily complicated re-description of what we are trying to explain, that is because we've reached a point where we cannot explain the phenomenon in any further terms. For Strawson, the answer to a question like "why does Jones deserve censure?" would be "Because he demonstrated ill will." If we then ask "Why do those who demonstrate ill will deserve censure?" Strawson's reply would be that we've reached a point where our justifications and explanations run out; we can only point to the practice and describe how it works. That those who demonstrate ill will deserve censure is an instance of what Wittgenstein called a "hinge" proposition: a proposition that can justify other propositions but cannot itself be justified by anything else.[21]

Perhaps it will be conceded that *if* a demand is legitimate, then certain reactions to the failure to meet the demand are appropriate. So when we wonder about appropriateness we aren't wondering why legitimate demands make certain reactions appropriate, but rather whether any demands are in fact legitimate. There are a number of ways we might understand challenges to the legitimacy of a demand. One way is to challenge the *content* of the demand. For example, suppose that an employer condemns one of her employees on the grounds that the employee did not flatter the employer. In this case a specific demand is in question and we can imagine one giving reasons for why flattery isn't a legitimate demand. We might appeal to general considerations of integrity or modesty or equity to argue against the demand. Here we appeal to other demands to challenge the demand for flattery. The legitimacy of the demand is considered in a context in which demands as such are not called into question but rather are considered in relation to each other. Because this kind of challenge is presented *within* the participant perspective, it assumes that at least some demands are legitimate.

In the above example, our challenge to the legitimacy of the demand was directed at the content of the demand, but as Strawson himself recognizes, there are circumstances in which no demands can be legitimately made of others. For a demand to be legitimate, those of whom the demand is made must possess the capacities necessary for establishing interpersonal relationships or entering into moral communities. Small children and nonhuman animals lack such capacities, so resentment, condemnation, praise, and gratitude are not appropriate responses to their behavior. Insofar as mentally deranged human beings also lack such capacities, the demand for goodwill cannot be legitimately made of them either. In these cases, we abandon the participant perspective and adopt what Strawson calls an objective attitude toward others. When we adopt an objective attitude toward someone, we see this being "as an object of social policy; as a subject for what, in a wide range of sense, might be called treatment; as something certainly to be taken account, perhaps precautionary account, of; to be managed or handled or cured or trained."[22]

When we abandon the participant perspective toward a human being, we no longer view this human being as a member of the moral community, but while Strawson thinks we often have good reasons to abandon the participant perspective toward other human beings, he does not think those reasons license the wholesale rejection of the participant perspective. As we saw in Chapter 2, however, there are those—including Greene and Eagleman—who think the truth of determinism does provide such license. In terms of the goodwill demand, we can put the argument as follows: the goodwill demand is legitimate only if a person can meet the demand—*ought implies can*. But if determinism is true, then people who fail to meet the goodwill demand cannot meet this demand—they "could not have done otherwise" than fail. Thus the goodwill demand is not legitimate, and those who fail to meet it are not responsible and do not deserve condemnation and punishment. If successful, such an argument would seem to imply that an objective viewpoint is the only viewpoint we are entitled to adopt—recall Eagleman's claim that we may eventually think about bad decision-making in the same way we think about diabetes or lung disease.

If responsibility and desert have their home within the participant perspective only, then Greene and Eagleman counsel abandoning the participant perspective. Remember, however, that Greene and Eagleman do not think such abandonment would leave us without a justification for reward and punishment. Reward and punishment can be justified on consequentialist grounds, e.g., we might justify reward and punishment on the grounds that reward and punishment are effective means for regulating behavior in socially desirable ways. Strawson anticipates this position, what Strawson calls the position of the *optimist*. The optimist is committed to two theses: (1) determinism rules out desert and responsibility; (2) reward and punishment can be justified on consequentialist grounds. Strawson criticizes both theses,[23] but while his criticism of (1) is the more contentious and widely discussed of the two, I am going to focus on his criticism of (2).

Strawson's primary challenge to thesis (2) is that the efficacy of rewarding and punishing that the optimist appeals to in justifying those practices depends on the very attitudes the optimist says we cannot justify. Strawson writes:

> It is far from wrong to emphasize the efficacy of all those practices which express or manifest our moral attitudes, in regulating behavior in ways considered desirable … . What *is* wrong is to forget that these practices, and their reception, the reactions to them, really *are* expressions of our moral attitudes and not merely devices we calculatingly employ for regulative purposes. Our practices do not merely exploit our natures, they express them. Indeed the very understanding of the kind of efficacy these expressions of our attitudes have turns on our remembering this.[24]

Though Strawson does not elaborate on this point, I think we can reveal the kind of understanding Strawson has in mind here by reflecting on two different ways we might conceive of the role the reactive attitudes play in regulating behavior. The optimist may be willing to concede that the efficacy of reward and punishment depends on the attitudes, but it is available to the optimist to claim that though the attitudes may be necessary to achieve the intended effects, we do not have to view them as appropriate, in a sense of "appropriate" that would provide a nonconsequentialist justification of the concepts of desert and responsibility. It is a familiar tactic, one in which we acknowledge the practical utility of a set of practices or beliefs without thereby seeing them as justified in any further sense. On this picture, we might say that although no one *actually* deserves punishment and no one is *actually* responsible for their actions, it turns out that thinking and acting *as if* people deserve punishment and *as if* they are responsible leads to desirable outcomes. For example, Patricia Churchland writes:

> If the only known way for "social decency" circuits to develop requires that the subject generate the relevant feelings pursuant to social pattern recognition, and if that, in turn, requires experiences of praise (pleasure) and blame (pain) consequent upon his actions, then treating the agent as responsible for behavior is a pragmatically necessary operating assumption. That is, it is justified by practical necessity.[25]

Oliver Goodenough provides another example. In the context of discussing the goals of punishment, Goodenough writes:

> I believe that these goals lead to a commitment … to treating the other agent as if he/she had the capacity to fully integrate the threat of punishment into its decision-making calculus, and to act accordingly, i.e. as if he/she had a kind of free will. Declaring this committed position both neutralizes attempts at deception by the transgressor and to some degree forces the role of a considering agent on the other player. However counterfactual the free will proposition may be in a deterministic world … it is a strategic fiction that underlies the productivity of a punishment rule, and is a fiction that may be deeply lodged in human cognitive and emotional psychology. Our free will intuitions may be false in the world of deterministic science and yet nonetheless effective in the world of strategic interaction.[26]

Here, responsibility—and by extension, desert and the moral reactive attitudes—is conceived merely as an effective instrument for achieving desirable outcomes, e.g., D-responsible agents.

What this picture leaves out, however, is the way in which responsibility and desert make our practices *intelligible* to us. When we take desert, responsibility, and the reactive attitudes seriously—when we credit them—what emerges are moral agents who become D-responsible in virtue of the fact that the demand for goodwill constitutes a reason for action. I am responsible for my failures to meet the goodwill demand, and censure and remorse are appropriate responses to these failures. Because I am responsible for these failures, there is something I can do to make sure I meet the demand in the future, and because I recognize the legitimacy of the goodwill demand, I have a reason to take steps to ensure that I meet the demand. D-responsibility emerges as a normative ideal implied by the goodwill demand, and the resulting D-responsibility is explained by the fact that there is an internal connection between D-responsibility and one's reasons for action. The practice of seeking forgiveness provides a good illustration of the kind of intelligibility in play here, the kind in which there is an internal connection between D-responsibility and one's reasons for action. Strawson notes that to seek forgiveness "is in part to acknowledge that the attitude displayed in our actions was such as might properly be resented and in part to repudiate that attitude for the future."[27] In repudiating an attitude for the future, the person who seeks forgiveness signals that though she failed to meet the goodwill demand, it is within her power to meet the demand. But in contexts in which forgiveness is sought, having the power to meet the demand is not enough. If I am the offended party and am considering whether to forgive the offender, I want to know whether the offender will *exercise* her power to meet the demand. Assuming that forgiveness is sometimes reasonably granted, sometimes we are justified in believing that offenders will exercise their power to meet the goodwill demand. What could provide such justification? In large part, the justification comes from the fact that we view the offender as a member of the moral community for whom the goodwill demand provides reasons for action. For example, suppose that a friend fails to show up for a lunch date, a failure that results from her simply forgetting about it. Suppose further that she knows that she would not have forgotten about it had she put the lunch date on her calendar. If your friend recognizes the legitimacy of the goodwill demand, then the demand will provide her a reason to be more diligent about writing down appointments in the future. But in seeing herself as having a reason to be more diligent, your friend thereby sees that acting on this reason will make her more reliable at keeping appointments, that is, bring her closer to being a D-responsible person.

With these points in mind, let's return to the example of Spiderman. Recall that I said it is the nature of his reaction that lends credence to the claim that a legend is born and that it can lend such credence because his reaction makes it more likely that he will exercise his power in future situations. We are now in a better position to see why. Spiderman's reaction reveals to us that he is someone who recognizes the legitimacy of certain demands and views failures to meet these demands as *his* failures. In seeing him this way, we are bound to see him as someone for whom certain demands function as reasons, reasons that are internally related to becoming a D-responsible person. *That* is why we believe a legend is born.

The intelligibility provided by desert and responsibility, i.e., the intelligibility that emerges when we view ourselves as moral agents who act for reasons, is what gives legitimacy to the goodwill demand. Put another way, what makes the goodwill demand legitimate is simply *the fact* that the goodwill demand can function as a reason for action in a way that is internally related to D-responsibility. We cannot legitimately demand of children, nonhuman animals, and the mentally deranged that they express goodwill toward us. We cannot legitimately demand this because the demand for goodwill cannot provide them a reason for action. In appreciating these points, we expose the emptiness of the distinction between "actual" desert and responsibility and "as if" desert and responsibility. Suppose the optimist grants that understanding the efficacy of our practices requires the kind of intelligibility I've described but argues that "as if" desert and responsibility provides it, i.e., argues that "as if" desert and responsibility is sufficient to explain the way in which the goodwill demand can function as a reason for action that is internally related to D-responsibility. If the optimist concedes that serving this function is what explains the efficacy of our practices, the optimist no longer has any grounds for denying "actual" desert and "actual" responsibility; after all, the optimist's rejection of desert and responsibility turned on the claim that the goodwill demand is not legitimate, but the legitimacy of the demand turns on the very intelligibility that the optimist now acknowledges.

THE PRIMACY OF A-RESPONSIBILITY

I said that moral responsibility can be understood in terms of P-. D-, and A-responsibility, but as I've characterized them, P-, D-, and A-responsibility are applicable regardless of whether we adopt the participant perspective toward other people, i.e., we can make sense of these concepts of responsibility even if we adopt an objective viewpoint toward humans. If moral responsibility has its home within the participant perspective only, we will need to refine one or more of these three types of responsibility. As my discussion to this point suggests, what distinguishes the participant perspective from the objective viewpoint (the viewpoint of the optimist) is how A-responsibility is conceived. P-, D-, and A-responsibility each reflects what I will call the *cognitive-affective architecture* (CAA) of a person. Broadly speaking, a person's CAA has to do with the way a person thinks and feels, including how a person reacts to situations (e.g., what they are prompted to think, what they are prompted to feel), as well as how a person's thoughts, feelings, and attitudes interact to produce behavior. So understood, one's CAA includes values, preferences, and personality traits, but it also includes a wide range of skills and capacities related to memory, awareness, reasoning, learning, and perception. P- and D-responsibility each concerns a person's CAA considered at a particular moment in time. By contrast, A-responsibility concerns the potential to change a person's CAA over time. As should become clear in what follows, the kinds of changes to a person's CAA that are relevant to A-responsibility can range anywhere from fairly large alterations in temperament to relatively small modifications in attention.

Minimally, to say that a person is A-responsible is to say that his CAA can be altered. But it is not enough merely to be subject to alteration. To be A-responsible, a person must have a CAA such that alterations in the CAA can result in the person being D-responsible. Given that the optimist hopes to modify behavior in socially desirable ways, she will accept all of this. Eagleman suggests that we might come to view bad decision-making as any other physical process, like diabetes or lung disease, and to the extent that the latter can be treated medically, bad decision-making might be so treated as well. This idea is already with us, of course; think of the now routine use of prescription medications to manage various mood and attention deficit disorders. We can view these medications as altering a person's CAA, and since decision-making is a function of one's CAA, there is no in principle reason to believe that bad decision-making cannot be similarly managed—perhaps with drugs, hypnosis, or surgery. When we think about altering a person's CAA this way, we adopt an objective viewpoint toward A-responsibility. By contrast, the participant perspective on A-responsibility rests on the idea that, since certain demands, e.g., the goodwill demand, can function as reasons for action, such demands can initiate changes in a person's CAA. As we've seen, whether such demands will initiate changes in a person's CAA depends on whether the person views herself as P-responsible for failing to meet such demands and whether she reacts to such failures appropriately. The point is nicely illustrated by Daniel Dennett, who discusses the case of a father whose child died after he inadvertently left the child alone in a car for several hours. Dennett says of the father that "if he makes himself really small, he can externalize this whole episode in his life, almost turning it into a bad dream, a thing that happened to him, not something he did. Or he can make himself large, and then face the much more demanding task of constructing a future self that has this terrible act of omission in its biography."[28] Even if we believe that there is a fact of the matter about whether this episode is something that happened to the father or something he did, a fact that does not depend on what the father thinks about the episode, Dennett's point is that how the father thinks about the episode is relevant to his future self. Like Spiderman, what the father does next time can depend on whether he sees this tragic episode as his fault and remembers it as such. Remembering *this happened because of my inattentiveness* can have the effect of making the father more attentive in the future and thus serve as the basis for a pledge that it won't happen again. Part of what makes this task "more demanding" is that whether the father remembers the episode plausibly depends on whether he reacts with guilt, shame, or remorse. As anyone who has experienced them knows, these self-directed attitudes are painful (or at least unpleasant), and it may be precisely this fact about them that explains why we remember our failures. Nietzsche puts the point bluntly:

> "How can one create a memory for the human animal? How can one impress something upon this partly obtuse, partly flighty mind, attuned only to the passing moment, in such a way that is will stay there?" … "If something is to stay in the memory it must be burned in: only that which never ceases to *hurt* stays in the memory"—this is a main clause of the oldest (unhappily also the most enduring) psychology on earth.[29]

Hyperbole aside, we should not ignore the fact that there has to be some means by which a person's CAA is changed. The goodwill demand can serve as a reason for action, but understanding how this can have any practical bite requires some way of relating reasons for action to an agent's psychology. In short, we need to understand how the goodwill demand can become an *effective* reason.[30] As responses to the failure to meet the goodwill demand, shame and remorse are appropriate, in the sense of merited or deserved. But they are also psychologically relevant in the sense that they are part of the mechanism by which changes in one's CAA are made. In being appropriate responses to failure, these attitudes signal the right of others to make the demand; in being part of the mechanism that transforms the goodwill demand into an effective reason, these attitudes help ensure that this right is honored. Thus we can see that the goodwill demand plays an explanatory role even in those cases in which the goodwill demand is not met. Spiderman did not meet the demand, but insofar as his CAA is structured so as to bring about changes that make it more likely that he will meet that demand in the future, we can say that one function of his CAA is to produce D-responsibility.[31]

If P-responsibility is the precondition for moral responsibility and D-responsibility is its end, the *drama* of moral responsibility—including our practices of praising and blaming, rewarding and punishing—has its stage in A-responsibility. When we adopt the participant perspective toward others, we see them as moral agents for whom certain demands play an important role in regulating thought and behavior. Moral agents can fail to meet the demands made of them, but what is most crucial to being a moral agent is the way in which demands can function as reasons for action. Recall the earlier claim that only D-responsible people have the right to make promises, or that only D-responsible people can make themselves into a pledge for the future. We can now see that this isn't quite true. Some *not yet* D-responsible agents can make themselves into a pledge for the future because some of these agents are A-responsible. A person who takes responsibility for failures pledges a D-responsible future self, and if such a person is A-responsible, it is her right.[32]

Eagleman's suggestion that we may view bad decision-making like any other physical process, e.g., diabetes or lung disease, is an example of adopting an objective viewpoint toward humans. This may give the impression that the participant perspective is committed to the claim that bad (or good) decision-making is a nonphysical process, but this impression would be mistaken. Eagleman characterizes free will in a way that makes free will something that operates independent of the rules of biology, so he thinks that if all behavior has a wholly biological basis, then none of our behavior is the result of free will; by implication, we are not morally responsible for our actions. As I've characterized them, the critical difference between the participant and objective viewpoints has to do with whether changes in a person's CAA can be initiated by normative demands, and as I have characterized moral responsibility, to be morally responsible is to be A-responsible in the way described by the participant viewpoint, i.e., to be such that normative demands can initiate changes in one's CAA. This implies dualism only if it is not possible for a wholly physical system to have this kind of CAA, but this impossibility has not yet been shown. What those who want to maintain the integrity of the manifest image—as embodied in the participant perspective—while rejecting dualism need to say is that moral responsibility is a property or feature of a subclass of physical entities or systems. The criticism of the objective viewpoint is *not* that it treats human beings as wholly physical systems but rather that is fails to honor important distinctions *within* the class of physical systems.

Missing from the optimist's outlook on humans is the intelligibility responsibility and desert bring to our practices of rewarding and punishing, and the optimist is wrong that finding room for this kind of intelligibility requires rejecting determinism. But the optimist is right about two important matters. First, if rewarding and punishing are means by which D-responsible agents are produced, it is available to the optimist to argue that there are more effective means for achieving this end, e.g., surgery or drugs. In effect, the optimist can concede that desert and responsibility are compatible with determinism but then point out that reward and punishment are crude instruments for achieving D-responsibility. While this might not entail the complete rejection of these practices—and all that comes with them, including responsibility, desert, and the reactive attitudes—we might come to appreciate their limitations and recognize the need to augment them. Showing that responsibility and desert are compatible with determinism is not the same as showing that the participant perspective is privileged, and while it may seem to many that the use of drugs or surgery would be dehumanizing, we shouldn't fetishize the participant perspective. This leads to the second, related way in which the optimist is right. Demonstrating the compatibility of responsibility and determinism is not the same as demonstrating that anyone is in fact responsible, and even if we accept that some people are responsible, it may turn out that the class of responsible agents is much smaller than we take it to be. Crediting the outlook of the optimist may thus provide us with an antidote to what Michael Perlin calls "sanism,"[33] which is the tendency to treat a person as sane even if there is powerful evidence to the contrary. Those who, like Eagleman, think that neuroscience is poised to have a significant impact on questions of responsibility are right, but not because neuroscience will convince us of determinism. Rather, the impact concerns the potential of neuroscience to reveal limitations on the capacities necessary for membership in the moral community. In this chapter I've indicated what these capacities are, and over the next few chapters I'll have more to say about them. Once we get clear about what responsibility requires, we will be in a better position to assess the limitations of the participant perspective.

REVISITING MR. PUPPET

Strawson thinks that the issue of determinism is irrelevant to the practice of holding people responsible, but as we've seen, Greene and Cohen think that thought experiments like the one involving Mr. Puppet show that assumptions about determinism do play a role in our judgments about responsibility. So who is right—Strawson, or Greene and Cohen? Before weighing in on this matter, it is important to keep separate two, albeit related, issues. The first issue is whether Mr. Puppet is responsible for his crimes. The second is what should happen to Mr. Puppet if he is responsible. Even if we assume that Mr. Puppet deserves some jail time and we think he should go to prison for life, it does not necessarily follow that he deserves to go to prison for life. The reason is that questions of responsibility and desert are not the only questions relevant to imprisonment. For example, given the nature of Mr. Puppet's crime, we may judge that it is too risky to let him reenter society. Justifications for punishment—or better, incarceration—can outstrip desert, a point we will return to in Chapter 7.

The success of the thought experiment requires more than the judgment that Mr. Puppet is not responsible for the crime. We must also judge that (a) no one is responsible for the crime, and (b) it is the truth of determinism *per se* that explains why Mr. Puppet is not responsible. For example, if you concluded that the scientists were responsible for the crime, then you missed the point of the thought experiment. Moreover, if you concluded that Mr. Puppet is not responsible on the grounds that he was manipulated by other agents, then you again missed the point of the thought experiment. Regarding this last point, compatibilists and incompatibilists alike recognize there are situations in which an agent is not responsible for her actions, situations that do not assume the truth or falsity of determinism. So, for example, suppose Mr. Puppet was so mentally deranged that he believed he was the King of France and had a head made of cabbage. In such a case we might very well judge that he is not responsible for his crimes, regardless of whether or not we think determinism is true.

Does the thought experiment give us clear grounds for accepting both (a) and (b)? There are good reasons to doubt that it does. With respect to (a), recall the testimony provided by the lead scientist. The scientist says that he designed Mr. Puppet "so that he would become precisely what he is today," and adds at the end, "I assure you that the accused deserves none of the credit." That the accused deserves none of the credit is consistent with both *no one* deserving the credit and *someone else* deserving the credit, but the latter is the much more intuitively plausible of the two. Imagine that the scientist is asked a follow up question: "Does anyone deserve credit?" We know his answer: "*I* do!" Our willingness to grant that Mr. Puppet is not responsible seems to turn precisely on the fact that there is someone else on whom we can place responsibility. Moreover, we come to see a basis for doubting (b) when we reflect on our reason for holding the scientist responsible. Mr. Puppet is a *puppet*, and in viewing him as such, we see his actions as having their source in someone other than him. It is this fact, not the truth of determinism per se, that explains why we do not view Mr. Puppet responsible for his crimes.

It might be objected that Mr. Puppet is not a puppet in the customary sense of a puppet; after all, the customary sense of a puppet is someone whose behavior is generated in a way that bypasses their own decision-making processes, but this is not true of Mr. Puppet—the scientists do not bring about his behavior directly, but indirectly through their design of Mr. Puppet's decision-making processes. As such, the fact that he is not responsible cannot be explained in the same way that we would explain why a customary puppet is not responsible. It is determinism per se that explains why Mr. Puppet is not responsible. Such an objection would be misplaced in the present context, however, for in denying that Mr. Puppet is a customary puppet, we run the risk of retracting the grounds for seeing him as not responsible in the first place. Put another way, Greene and Cohen need to provide justification for calling the man *Mr. Puppet* while at the same time denying that his actions can be excused in the way that the behavior of customary puppets would be excused, but they have not provided such justification. I submit that Greene and Cohen's thought experiment capitalizes on the customary sense of "puppet" while disavowing the very elements that explain our intuitions in such cases. We should say either Mr. Puppet is a customary puppet or he is not. If he is, then though he is not responsible for his actions, his lack of responsibility does not turn on the truth of determinism per se. If he is not, then it is simply tendentious to call him "Mr. Puppet," and the assertion that he is not responsible turns merely on the claim that determinism rules out responsibility—the very claim that the thought experiment is meant to establish.[34]

This potentially misleading feature of the thought experiment can be removed, of course. Greene and Cohen don't believe that there are teams of scientists designing criminals, but they also do not think it matters. If we replace designing agents with mindless genetics and environmental conditioning—mindless in the sense that these causes operate without any goal of creating a criminal—we still end up with a person whose crime is the result of forces beyond his control. Mr. Puppet is not responsible, and no one else is either. A number of considerations count against this conclusion, however. Consider what Greene and Cohen would have to say about Spiderman. Spiderman acquired his great powers as the result of being bit by a mutant spider. Let's stipulate that his getting bit was beyond Spiderman's control. If Greene and Cohen are right, then Spiderman bears no responsibility for Ben's death. Parity of reason would require Greene and Cohen to add that, upon Spiderman's transformation into a legend, "he is, perhaps, a man to be welcomed, and we would want him in the streets actively fighting crime. But given the fact that forces beyond his control played a dominant role in causing him to rid the city of crime and injustice, it is hard to think of him as anything more than a pawn."[35] Though we tend to think of responsibility in the context of our failures, the notion applies to our successes as well. If it is unfair to hold Mr. Puppet responsible for his crimes, it is equally unfair to view Spiderman as responsible for his acts of heroism. If the implication is that Mr. Puppet does not *deserve* blame, it also follows that Spiderman does not deserve praise. Imagine that the city council is considering whether to hold a parade in honor of Spiderman. If Greene and Cohen were on the council, they would have to argue that because it was beyond Spiderman's control whether he acquired those powers, Spiderman deserves none of the credit and therefore is not *deserving* of a parade in his honor. Of course, Spiderman differs from Mr. Puppet in that we are not imagining that the mutant spider planned to bring about a superhero, but it is crucial to keep in mind that Greene and Cohen do not think that this difference is relevant to the argument. *My* intuition is that we would not be persuaded that Spiderman does not deserve the parade. Spiderman did not ask for these powers, of course, but so what? He now has them, he knows he has them, he knows he can use them to rid the city of crime and injustice, and because he thinks ridding the city of crime and injustice is a worthy aim, he does so. Insofar as we understand the concept of *deserving praise* at all, it is hard to imagine a clearer case where the concept applies.

A second consideration concerns our intuitions about variations on Mr. Puppet-type cases, variations that hold fixed the truth of determinism. First, we get rid of the group of scientists and replace them with the complex, but deterministic, interaction of genes and environment. Second, instead of imagining a murderer, let's imagine a person who steals some groceries from a local convenience store. Let's call her Ms. Demeanor. At her trial, Ms. Demeanor expresses remorse and accepts blame, and she vows never to do it again. She spends six months in county jail, keeps her vow upon release, and spends the rest of her days as an upstanding member of the community. The law held Ms. Demeanor responsible for her theft and she took responsibility for it, but Greene and Cohen would have to say that *this was unfair*. Of course, Greene and Cohen might concede that our intuition of unfairness is less strong than in the case of Mr. Puppet, perhaps on the grounds that the consequences for Ms. Demeanor are less severe than the assumed consequences for Mr. Puppet. But such a reply would suggest that Ms. Demeanor's case differs from Mr. Puppet's only in being a more palatable injustice, and this, too, seems false.

So is Mr. Puppet responsible or not? My aim has been to show that it isn't obvious that the truth of determinism precludes him from being responsible, and I'd be happy for a draw at this point. I'll have more to say about Mr. Puppet later, and as I hope to show, whether Mr. Puppet is responsible turns on questions left unanswered in the original description of his case. But enough for now.

COULD HAVE CHOSEN OTHERWISE

The traditional compatibilist says that the truth of "I could have stopped Ben's death" turns on the truth of a counterfactual statement about what would have happened if the original circumstance had been different. For compatibilists of this stripe, the relevant differences we are to image are typically differences concerning features of the agent's wants, beliefs, intentions, or efforts, holding everything else about the circumstance as much the same as possible. In Spiderman's case, the relevant counterfactual might be something like "I would have stopped Ben's death if I had tried to prevent it," and, of course, we are meant to believe that this is in fact true. In revealing that Ben's death depended on facts about Spiderman's wants, beliefs, intentions, or efforts, the truth of the counterfactual reveals how the death depended on *him* and thus why Spiderman is P-responsible for the death. In the last chapter we considered the objection that this counterfactual is relevant only if it is also true that *whether Spiderman tried* depended on him, and the problem was that this sets off a regress of counterfactuals that reaches a point where what Spiderman "could have done" no longer turns on factors that depend on *him*. The objection evidently rests on the assumption that if there are factors whose existence do not depend on Spiderman and these factors are sufficient to bring it about that Spiderman does not save Ben then Ben's death does not depend on Spiderman. But this objection betrays a misunderstanding of *why* the truth of the original counterfactual is important, a point that can be made clear when we consider the relevance of the counterfactual to the appropriateness of *taking* responsibility.[36] As already discussed, whether it is appropriate for an agent to take responsibility depends in part on whether the agent possesses the capacity to adapt future behavior in goal-oriented ways. In the last chapter we discussed Tourette's patients, and let's imagine a situation in which someone with Tourette's blurts out an offensive remark, say at a dinner party. Would it be appropriate for this person to take responsibility for making the offensive remark? Put another way, can we legitimately demand that this person refrain from making offensive remarks? To the extent that the remark did not depend on the person's beliefs, wants, intentions, or efforts, changes in those beliefs, wants, intentions, or efforts will not bring her future behavior into compliance with the demand in question. The demand cannot provide her a reason to do anything, or at least not anything that would bring her closer to D-responsibility. As such, it would be inappropriate for the person to *take* responsibility. Now we see the obvious relevance of the issue of whether the person "could have chosen otherwise," understood as a question about what would have happened if her beliefs, wants, intentions, or efforts had been different. Contexts in which a person takes responsibility for a failure are contexts in which a person is prepared to seek forgiveness, to turn herself into a pledge for the future. This implies the exercise of certain powers, and whether the agent possesses those powers turns precisely on the truth of counterfactuals about what would have happened had the agent's beliefs, wants, intentions, or efforts been different. Lacking such powers, there can be no question about their future exercise; possessing such powers, the demand can provide an effective reason for bringing about their appropriate exercise.

But suppose we now ask "Could Spiderman have tried to prevent Ben's death?" We are going to have to block the regress at some point, so let's stop it here. Consider a variation on the Frankfurt case discussed in the last chapter. Imagine that a man named Black has inserted a chip into Spiderman's brain such that if Spiderman starts to try to save Ben, Black will activate the chip and quell Spiderman's effort. Let's also suppose this constitutes a situation in which Spiderman could not try to save Ben. Does it thereby follow that this constitutes a situation in which Spiderman could not save Ben? The traditional compatibilist will say no, on the grounds that what is relevant to whether Spiderman could have saved Ben is whether he would have saved Ben if he had tried, *not* whether he could have tried. But if we accept that Spiderman could not have tried to save Ben, it's hard to see how saving Ben constituted a genuine possibility. So what should we say here?

When Frankfurt challenged the claim that an agent is morally responsible for what she does only if the agent could have chosen (or done) otherwise, he took himself to be challenging what he called the principle of alternate possibilities. I have just said that in the variation of the Frankfurt case involving Spiderman, it is hard to see how saving Ben constituted a genuine possibility given that Spiderman could not have tried to save him. What this suggests is that though the traditional compatibilist's reading of could have chosen (or done) otherwise can generate true counterfactuals, it doesn't generate alternate possibilities. However, I've also argued that the traditional compatibilist's reading of could have chosen (or done) otherwise is crucial to understanding the phenomenon of responsibility. We should therefore conclude that, as far as responsibility goes, alternate possibilities are irrelevant, but we should resist concluding from this that the traditional compatibilist's reading of could have chosen chosen (or done) otherwise is thereby irrelevant to responsibility. To put the point another way: if we use the phrase "could have chosen (or done) otherwise" to express the principle of alternate possibilities and become convinced that a person can be morally responsible for what she does even if the principle of alternate possibilities is false, it will be awfully tempting to conclude that whether a person could have chosen (or done) otherwise is not relevant to whether the person is morally responsible. But this is a mistake, particularly when it comes to understanding the phenomenon of taking responsibility. To repeat a now familiar theme, if being responsible for Ben's death simply means being P-responsible for Ben's death, then being responsible implies nothing about whether Spiderman's behavior will be different next time. But taking responsibility is not just a matter of judging oneself P-responsible; it also has to do with how that judgment affects the person's CAA, i.e., A-responsibility is the mechanism that enables D-responsibility to emerge from P-responsibility. Whether Spiderman can meaningfully take responsibility for Ben's death thus depends on whether taking responsibility will increase the probability that the conditions specified in the counterfactual—i.e., the conditions that, if met, would have resulted in saving Ben—will obtain in relevantly similar future situations. If the counterfactual is true, and Spiderman knows this, Spiderman has a reason for action aimed at D-responsibility. *This* is why the counterfactual is so important.

Take another look at the cartoon panel with which I started this chapter. It is fair to say that the recent philosophical literature on responsibility has focused primarily on the left side of the panel: is Spiderman responsible for Ben's death? If so, does this depend on whether he could have prevented Ben's death? If so, what is the analysis of "could have" that is relevant to his being responsible? These are important questions, to be sure, but we should not consider the left side of the panel in isolation from the right. Each side tells us something about the phenomenon of moral responsibility, but understanding moral responsibility also requires us to understand the relationship between the panels, the way they make sense together. Understanding this relationship requires us to examine the practice of taking responsibility, and I've argued that this practice rests on the concept of A-responsibility. The significance of taking responsibility is clear in our interpersonal relationships, but it is equally significant to law. Our interest in legal responsibility is ultimately an interest in *holding* people legally responsible. We want to do justice to this practice, and while we might not care whether the people we hold responsible actually take responsibility, a justification of the practice of holding people responsible should also justify the practice of *taking* responsibility. When we get clear about what is involved in taking responsibility, we can also get clear about the conditions under which taking responsibility does or does not make sense.

NOTES

1. "Hurricanes in History," National Hurricane Center, accessed October 1, 2016, http://www.nhc.noaa.gov/outreach/history/#new.

2. Michael Levin, "A Compatibilist Defense of Moral Responsibility," in *Introduction to Philosophy*, ed. Louis Pojman and James Fieser (New York: Oxford University Press, 2008), 456.

3. To be clear, to say that Spiderman is responsible for Ben's murder isn't to say that Spiderman murdered Ben, and since Ben was murdered, we can say the murderer is (or potentially is) also responsible for Ben's murder.

4. For a related treatment of the distinction between what I am calling P- responsibility and D-responsibility, see Nicole Vincent, "Responsibility: Distinguishing Virtue from Capacity," *Polish Journal of Philosophy* 3(1) (2009): 111–126.

5. Friedrich Nietzsche, *On the Genealogy of Morals*, trans. Walter Kaufmann and R.J. Hollingdale (New York: Vintage Books, 1989), 57.

6. Nietzsche, *Genealogy*, 58.

7. Nietzsche, *Genealogy*, 58.

8. Nietzsche, *Genealogy*, 58.

9. Nietzsche, *Genealogy*, 59–60.

10. In the context of this specific example, this seems like a reasonably safe interpretation. But I think there are cases where the occurrence of an event did not depend on how a person acted but in an important sense still had something to do with the person. For example, suppose a birth-of-Spiderman story wherein he just *reads* about the murder of a man *like* Uncle Ben. He was in no position to have done anything about *that* murder, but he also recognizes that he has the ability to act in ways that might prevent future crimes of that sort against that sort of victim. In this case there's no question that the murder "didn't depend on how he acted," but he still might think it "has something to do with him" in a closely related sense: he cares about Ben (and, to a presumably lesser extent, innocent citizens like Ben), and he can see how he *could* act (or fail to act) in ways that might prevent future crimes of that sort. The point here is that you could get D-responsibility emerging without *actual* P-responsibility but rather from a sense of one's own "counterfactual" or "hypothetical" P-responsibility, i.e., P-responsibility not for a thing that actually depended on a person, but a projected or imagined sense of what kinds of things a person might have the power to affect. Indeed, I suspect that it is precisely this possibility that explains (at least in part) the way reading works of fiction can influence our moral development. Thanks to Ty Fagan for bringing this point to my attention.

11. Wilfrid Sellars, "Fatalism and Determinism," in *Freedom and Determinism*, ed. Keith Lehrer (New York: Random House, 1966), 145. As I will argue in Chapters 5 and 6, the conceptual irreducibility of which Sellars speaks is compatible with fusing the scientific and manifest images.

12. P.F. Strawson, "Freedom and Resentment," in *Free Will*, ed. Gary Watson (Oxford: Oxford University Press, 2003).

13. Strawson, "Freedom and Resentment," 84.

14. Strawson, "Freedom and Resentment," 80.

15. Strawson, "Freedom and Resentment," 85.

16. Strawson, "Freedom and Resentment," 84.

17. Strawson, "Freedom and Resentment," 84–85.

18. Strawson, "Freedom and Resentment," 90.

19. Strawson, "Freedom and Resentment," 90–91.

20. Strawson, "Freedom and Resentment," 77.

21. Ludwig Wittgenstein, *On Certainty*, trans. Denis Paul and G.E.M. Anscombe (New York: Harper & Row, 1969). That hinge propositions themselves cannot be justified by anything else does not entail that they are immune from explanation. I return to this point in Chapter 7.

22. Strawson, "Freedom and Resentment," 79.

23. Strawson's view is often characterized as a merit-based form of compatibilism. More recent attempts to defend such a view can be found in Fischer and Ravizza, *Responsibility and Control*, and Michael McKenna, *Conversation and Responsibility* (New York: Oxford University Press, 2012).

24. Strawson, "Freedom and Resentment," 93.

25. Patricia Churchland, "Feeling Reasons," in *On the Contrary* (Cambridge, MA: MIT Press, 1998), 252.

26. Oliver Goodenough, "Responsibility and Punishment: Whose Mind? A Response," *Philosophical Transactions of the Royal Society of London. Series B: Biological Sciences* 359 (2004): 1807. While Goodenough puts the point in terms of free will, it is fairly clear that the significance of free will attributions is the way in which they can justify judgments of responsibility.

27. Strawson, "Freedom and Resentment," 76.

28. Daniel Dennett, *Freedom Evolves* (New York: Viking Press, 2003), 302.

29. Nietzsche, *Genealogy*, 60–61.

30. By 'effective reason' I mean something in the spirit of what Donald Davidson calls a primary reason. Donald Davidson, "Actions, Reasons, and Causes," *Journal of Philosophy* 60 (23) (1963): 685–700. I'll have more to say about effective reasons in the next chapter.

31. The sense of 'function' intended here will be made clear in Chapters 5–7.

32. One implication of my account is that whether it is appropriate to blame someone for her actions depends, in part, on whether it is reasonable to believe that the person blamed can and will (or will try) to bring about appropriate changes in her CAA. This requirement may initially seem too strong, but it is important to keep in mind that the kind of changes in a person's CAA that are relevant to A-responsibility can include rather subtle changes. Moreover, the plausibility of the requirement becomes clear when we reflect on our expectations about the future behavior of those we blame. Recall the example of the friend who failed to make a lunch date. I might blame my friend for her failure, and in so doing it is reasonable for me to expect that my friend will show up the next time we make a lunch date. But it is hard to see how this expectation can be reasonable if I do not also assume that my friend will do something to reduce the likelihood that she will miss future lunch dates. However, in assuming that my friend will do something to reduce the likelihood of missing future lunch dates, I need not imagine that my friend undergoes some radical change in temperament or personality—she doesn't have to suddenly become a different kind of *person*, as it were. It might be enough that she becomes more mindful about keeping a calendar or setting alarms.

33. Michael Perlin, "On 'Sanism,'" *Southern Methodist University Law Review* 46 (1992): 373–407.

34. If Mr. Puppet is not a puppet in the customary sense, it might be thought that the fact that his decision-making processes were designed by the scientists is enough to show that Mr. Puppet is not responsible. I've two replies. The first is that this response won't help if we are still able to place responsibility on the scientists. The second is that Greene and Cohen don't think that our decision-making processes are

the result of designers, so if determinism rules out responsibility, it better not do so in any way that depends on the presence of designers.

35. Recall what Greene and Cohen say of Mr. Puppet: "He is, perhaps, a man to be feared, and we would not want to return him to the streets. But given the fact that forces beyond his control played a dominant role in causing him to commit these crimes, it is hard to think of him as anything more than a pawn."

36. The assumption looks to be false in any case. As Simon Blackburn notes, although a thermostat cannot control the past and cannot control the way the past controls the present and future, a thermostat can the control the future (in respect of temperature, at least). *Think: A Compelling Introduction to Philosophy* (New York: Oxford University Press, 1999), 83.

Chapter 4

What A Stupid I Am!

The 1968 Masters golf tournament should have been decided in a play-off; American Bob Goalby and Argentinean Roberto De Vicenzo took the same number of strokes over the four-day tournament, and fewer than any other player. But in golf, one's score is not determined by the number of strokes taken, but the number of strokes recorded on one's official, signed scorecard. De Vicenzo's official scorecard contained an error, crediting him with one more stroke than he actually took. De Vicenzo failed to check his scorecard before signing it, and the result was that he lost the tournament. Upon learning of his mistake, De Vicenzo exclaimed "What a stupid I am!" Like Spiderman, De Vicenzo took responsibility for this failure, a failure he viewed as resulting from his *stupidity*. In Chapter 1 I said that we should aim for an account of responsibility that makes sense of the way questions about responsibility are often posed as questions about free will, on the one hand, and the way questions about responsibility are often posed as questions about rationality, on the other. Building on the work of the previous chapter, I argue that free will is a type of responsibility distinctive to rational agents. The ideal of rationality functions as a focal point for adjustments in decision making. The account of freedom I develop is also intimately linked with the notion of autonomy: an autonomous individual is self-ruling, and *self-regulating*. The regulation concerns both the understanding, in the sense of tending toward increased knowledge, and the will, in the sense of tending to conform one's will to what one knows.

RATIONALITY AND REASONS

Throughout the discussion to this point I've made a number of claims about rationality and reasons for action. For example, I've said a rational person chooses the action that the balance of reasons favors, and a responsible agent is one for whom certain demands function as effective reasons for action. While these claims may have been unproblematic in the contexts in which I made them, it is worth making explicit how I am using "rational" and "reasons" (along with related expressions, e.g., "rationality," and "having a reason"). My point of departure is the following remark by Aristotle, made in Book VI of *Nicomachean Ethics*:

> As assertion and denial are to thought, so pursuit and avoidance are to desire. Now virtue of character is a state that decides; and decision is a deliberative desire. If, then, the decision is excellent, the reason must be true and the desire correct, so that what reason asserts is what desire pursues. This, then, is thought and truth concerned with action. By contrast, when thought is concerned with study, not with action or production, its good or bad state consists simply in being true or false. For truth is the function of whatever thinks; but the function of what thinks about action is truth agreeing with correct desire.[1]

By "thought," Aristotle means the part of the human soul whose function is to arrive at truth, and thus while Aristotle distinguishes thought concerned with study and thought concerned with action, each kind of thought involves truth. The difference between study and action has to do with the kind of truth at which thought aims. Study is concerned with truths that are universal and necessary, and thus invariable. By contrast, action is concerned with truths that are particular and contingent, and thus variable. This distinction can perhaps be made clear when we consider the role that thought plays in choice. Suppose I desire to be wealthy. What steps can I take to satisfy this desire? The job of thought is to provide knowledge of the best means by which I may satisfy this desire. But the best means by which I may satisfy this desire depend on my specific circumstance and may not be the best means by which someone else may become wealthy. For example, given my talents, education, and opportunities, perhaps going to business school is the best means by which to become wealthy. For others, wealth might be best achieved by pursuing a career in music. It is in this sense that the truths at which thought aims are particular, contingent, and variable.

For Aristotle, the good cannot be defined in terms of desire, i.e., desiring X does not entail that X is good. Common sense supports this: a person may desire to take heroin, but we don't thereby conclude that taking heroin is good for the person. And while Aristotle can allow that a person might desire what is good without knowing that what she desires is good, Aristotle believes we can achieve knowledge of what the good is. It is not my aim, and it is beyond the scope of this book, to defend a general account of the good.[2] In what follows, I simply assume that such an account can be given, and I assume that such an account would not deviate too far from a conception of the good that counts such things as physical and mental health, close personal relationships, and meaningful work as among the things that make for a life well lived.

Modern discussions of practical reason often focus on what it is to have a reason for doing something, and it is common to distinguish *normative* reasons from *motivating* or *explanatory* reasons.[3] Following T. M. Scanlon, we can say that normative reasons are considerations that favor adopting a belief or course of action.[4] Suppose it is good to be healthy and eating salad promotes health. That eating salad promotes health would then be a reason to eat salad. So understood, normative reasons turn out to be facts or states of affairs, e.g., one reason to eat salad is the fact that eating salad promotes health.[5] By contrast, explanatory reasons are reasons that a person takes herself to have and which can prompt her to act. Reasons in this sense are often taken to be psychological states of a person, e.g., beliefs and desires. If explanatory reasons are psychological states and normative reasons are facts or states of affairs, then strictly speaking explanatory reasons are never normative reasons. However, explanatory reasons can be normative reasons in the derivative sense that an explanatory reason can have a normative reason as its content.[6] For example, I may believe that eating salad promotes health, and so if the fact that eating salad promotes health is a reason to eat salad, and if my belief that eating salad promotes health in part explains my salad eating, then we can say that the explanatory reason is also a normative reason. An account along these lines seems to be at play in Aristotle's account of practical reason. Let's assume that the output of practical reasoning is a belief about the best means for satisfying the person's desire. If the means are actions, then we can say the belief-desire pair constitutes a person's reason for action. According to Aristotle, a belief-desire pair such as this is an "efficient" cause of choice, and choice is an efficient cause of action.[7] Belief-desire pairs explain the agent's actions in virtue of causing them, and it is in this way that belief-desire pairs are explanatory reasons, or what I've called *effective* reasons. But Aristotle recognized that reasons can be explanatory in this sense without the explanatory reason being a normative reason. I may desire to be wealthy and believe that going to business school is the best means to achieve it, and this belief-desire pair might cause me to choose to go to business school. If this pair is also a normative reason, then my choice will be right. But in order for the pair to count as a normative reason, the "reasoning must be true and the desire right." So there are a number of ways a belief-desire pair can be explanatory but fail to provide a normative reason: (1) the belief about means is correct, but the desire is wrong, i.e., I desire the wrong things; (2) the desire is right, but the belief about means is incorrect; (3) the desire is wrong and the belief about means is incorrect. Note, too, that similar points apply to the reasoning involved in study: just as practical reasoning may fail to achieve knowledge of the best means by which a desire may be satisfied, theoretical reasoning may fail to achieve knowledge of universal, necessary truths, e.g., in attempting to provide a demonstration of a universal, necessary truth, I might make an invalid inference and arrive at a false conclusion. In such a case there will be an explanatory reason for my arriving at the conclusion I do, but the explanatory reason won't be a normative reason.

Explanatory reasons are psychological states, and an explanatory reason is a normative reason, in a derivative sense, just in case the content of the explanatory reason is a normative reason. But a psychological state can have a normative reason as its content even if the psychological state is not an explanatory reason. For example, I might perform an action, believe that performing that action is the best means to achieve something I desire, and yet not perform the action for that reason.[8] Imagine that I desire to be healthy, I believe that eating salad is the best means to satisfy that desire, and I eat a salad. But suppose I also desire to impress my date and believe that eating salad is the best way to do this. It might be that my desire to impress my date was the causally relevant desire, i.e., it was the desire that explains my action (along with the belief about means). However, even if my desire to be healthy does not explain my salad eating, it is the kind of state that is a potential cause of behavior. We can thus say that having a reason to X is being in a psychological state whose content is a normative reason to X, where the psychological state is a potential cause of action or belief. In referring to a person's reasons, I will mean that the person has reasons in the sense just characterized.[9]

An ideal rational agent is one whose explanatory—or effective—reasons are normative reasons. More accurately, an ideal rational agent is one whose explanatory reasons accord with her all-things-considered normative reasons. An explanatory reason for performing an action might be a normative reason and yet the action performed may not be the rational thing to do. Suppose I order soup because I desire something tasty and believe the soup is the tastiest option, and assume that being tasty is a consideration in favor of ordering soup (it is a normative reason). Still, there may be a host of considerations that favor ordering salad instead, and all things considered, salad may be the better option. In this case, I don't have an all-things-considered normative reason to order soup. In terms I've used previously, an ideal rational agent believes and acts in accordance with what the balance of normative reasons favors. Two clarifications are needed, however. First, one way in which the balance of normative reasons might favor ordering salad over soup is that ordering salad would satisfy more of the person's desires than ordering soup would, e.g., the salad may be less expensive than the soup and the person desires to save money. Even so, we needn't commit ourselves to saying that the desire to save money was in fact an explanatory reason for ordering salad. All that is required is that the balance of the person's normative reasons favors ordering salad and that at least one of those normative reasons is an explanatory reason for ordering salad. Second, it is possible that while the balance of the person's normative reasons favors salad, there are normative reasons for ordering soup of which the person is unaware (these reasons are not the content of any of her psychological states). If including those normative reasons results in the balance of normative reasons favoring soup, then it seems that ordering salad is not the rational thing to do. In light of this, we can say than an ideal rational agent is one who believes and acts in accordance with what the balance of her normative reasons favors, at least one of her normative reasons is an explanatory reason, and what the balance of her normative reasons favors does not conflict with what the balance of normative reasons favors.

FREE WILL AND THE IDEAL CARTESIAN WORLD

There is a long philosophical tradition of understanding free will in terms of rationality, a tradition that includes Plato, Spinoza, and Hegel, among others. The account I develop draws on a conception of free will articulated in Descartes' *Meditations on First Philosophy*. In Meditation Four, Descartes writes:

> Whereupon, regarding myself more closely, and considering what my errors are (which alone testify to the existence of imperfection in me), I observe that these depend on the concurrence of two causes, viz, the faculty of cognition, which I possess, and that of election or the power of free choice—in other words, the understanding and the will. For by the understanding alone, I [neither affirm nor deny anything but] merely apprehend *(percipio)* the ideas regarding which I may form a judgment; nor is any error, properly so called, found in it thus accurately taken.[10]

Regarding the will, Descartes adds:

> The power of will consists only in this, that we are able to do or not to do the same thing (that is, to affirm or deny, to pursue or shun it), or rather in this alone, that in affirming or denying, pursuing or shunning, what is proposed to us by the understanding, we so act that we are not conscious of being determined to a particular action by any external force. For, to the possession of freedom, it is not necessary that I be alike indifferent toward each of two contraries; but, on the contrary, the more I am inclined toward the one, whether because I clearly know that in it there is the reason of truth and goodness, or because God thus internally disposes my thought, the more freely do I choose and embrace it; and assuredly divine grace and natural knowledge, very far from diminishing liberty, rather

> augment and fortify it. But the indifference of which I am conscious when I am not impelled to one side rather than to another for want of a reason, is the lowest grade of liberty, and manifests defect or negation of knowledge rather than perfection of will; for if I always clearly knew what was true and good, I should never have any difficulty in determining what judgment I ought to come to, and what choice I ought to make, and I should thus be entirely free without ever being indifferent.[11]

Descartes identifies the will with the power of free choice. So understood, *all* acts of will are free, so the phrase "free will" is redundant. Descartes initially characterizes the power of will as the ability "to do or not to do the same thing," but there is a compatibilist and an incompatibilist reading of this formulation. On the compatibilist reading, though a person does a thing, the person would have not done the thing had circumstances been different. That is the sense in which the person has the ability to do or not do the same thing. On the incompatibilist reading, though a person does a thing, in the actual circumstances in which the person did the thing, it was possible to not do the thing. However, Descartes immediately provides a second formulation, one that seems to make the correct interpretation of "ability" moot. According to the second formulation, an action results from our will just in case when we act we are not conscious of being determined to act by any external force. The rationale for the second formulation becomes clear in the next line, where Descartes not only denies that the power of free choice requires indifference toward two contraries but also argues that the more inclined he is toward one contrary the more freely he chooses it. In fact, there are times when the inclination may be so strong that we cannot but judge the way we do:

> For example, when I lately considered whether aught really existed in the world, and found that because I considered this question, it very manifestly followed that I myself existed, I could not but judge that what I so clearly conceived was true, not that I was forced to this judgment by any external cause, but simply because great clearness of the understanding was succeeded by strong inclination in the will; and I believed this the more freely and spontaneously in proportion as I was less indifferent with respect to it.[12]

On the face of it, saying that he "could not but judge" conflicts with the formulation according to which the power of free choice consists in the ability to do or not to do the same thing. The compatibilist might argue here that the conflict is not genuine because Descartes is simply wrong about whether he could not but judge the way he does, i.e., the compatibilist might say that Descartes would not have judged as he did if he had not had a clear and distinct perception. Even if the compatibilist is wrong about this particular counterfactual, the compatibilist makes a fair point. As we will see, Descartes thinks the correct use of free will requires that we refrain from affirming that which is not clearly and distinctly perceived. Insofar as some people do use free will correctly, Descartes would have to concede that the compatibilist counterfactual is at least sometimes true of people. But nothing in Descartes' discussion suggests the claim that he could not have judged otherwise turns on how he would have judged if he had not had a clear and distinct perception, so it seems safe to assume that Descartes had the incompatibilist reading of "could not but judge" in mind. Be that as it may, Descartes' claim that he could not but judge the way he did does not conflict with the second formulation, for all that is required to satisfy the second formulation is that his judgment not be the result of some *external* force. It thus appears that while Descartes accepts the incompatibilist's reading of the ability "to do or not to do the same thing" and accepts that the fact that he could not but judge the way he did entails that he did not have the ability to do or not to do the same thing, he *denies* that the ability to do or not to do the same thing is necessary for free choice.

Because the second formulation is the one seemingly most relevant to free choice, it is important to clarify what is meant by saying that an action (or a choice, or a judgment) is or is not determined by an external force. Because he was a dualist, Descartes would claim that his faculties of understanding and will are faculties of his immaterial mind. External forces, then, would include anything not belonging to one's immaterial mind. Physical forces would thus count as external, as would forces exerted by other immaterial minds (e.g., God). However, if we reject dualism but maintain that some choices are free, then we cannot allow that *all* physical forces are external. For example, if the faculties of understanding and will supervene on the brain—if *you are your brain*—then we cannot say the brain is an *external* force. What sort of physical forces would count as external, then? Recall Jeffrey Rosen's claim that "someone who has the illusion of making a free and rational choice between soup and salad may be deluding himself, since the choice of salad over soup is ultimately predestined by forces hard-wired in his brain." Assume for the moment that "being hard-wired" means genetically determined, and assume that anything relevant to the "choice" between soup and salad depends only on that which is genetically determined. Let's assume, too, that your genes count as an external cause such that if your choice of salad over soup can be correctly described as the effect of your genes, your choice was forced by an external cause. Granting all of this, it would seem that Descartes himself would say that the choice of salad was not free. But what exactly is the intended contrast here? In Descartes' own example, Descartes characterizes the judgment that he exists as not resulting from an external cause but rather a strong inclination of the will that is prompted by a clear and distinct perception. When Descartes talks about external causes, he means that which does not belong to the operations of the understanding and the will. Let's call that which does belong to the operations of the understanding and the will "internal," and let's stipulate that the clear and distinct perception (an operation of the understanding) caused the judgment (an operation of the will), in whatever sense of "cause" that accounts for Descartes' insistence that he "could not but judge" as true that which he clearly and distinctly perceived. We can thus say that there was an internal cause of his judgment, namely the clear and distinct perception. Putting all of this together, we might say that one's genes determine the existence of an understanding and will such that clear and distinct perceptions cause the will to affirm those perceptions as true. In this case, are the affirmations caused by something external or internal? The proximate cause would be something internal, for clear and distinct perceptions are provided by the understanding. However, insofar as one's understanding and will are the effect of one's genes, there is an indirect sense in which something external (genes) causes the affirmation, for the genes determine the existence of an understanding and will that has a certain causal structure. Would the fact that your genes—something we are calling external—can be correctly described as a cause of your affirmation entail that the affirmation was not free? No. When Descartes wonders whether a particular effect is the result of external causes, he is wondering whether or not the effect has anything to do with the operations of the understanding and will. External causes are thought of as *competitors* to internal causes such that if an effect has an external cause, then the effect is not the result of the understanding and will. But one's genes are not in competition with the understanding and will in this sense, and so one's genes do not constitute the kind of external cause that would threaten free will.

For Descartes, the absence of external determining forces is sufficient for a choice to be free. However, Descartes also notes that there are different grades of freedom. The "lowest grade of liberty" is the liberty of indifference, a state that results from a "defect or negation of knowledge." The highest grade of liberty is found in situations where a person has clear knowledge of what is true or good, and as just noted, far from being indifferent in those cases, a person's will can be greatly inclined in one direction—even to the point where one "could not but judge" the way she did. Grades of free will are thus linked to rationality in two ways. One aspect of rationality concerns the work of the understanding, that power in us which can reveal what is true or good. If I choose in the absence of clear knowledge, then although my choice will be free, it is an inferior form of free will. Another aspect concerns the work of the will, that power in us which can affirm what is true or choose what is good. While Descartes himself does not consider it, it seems in principle possible that one might fail to affirm—or perhaps even deny—what the understanding clearly and distinctly perceives. Imagine a case in which

I deny what I clearly and distinctly perceive. Insofar as my denial is not the result of some external determining force, it will be free. But because such a denial would clearly constitute a failure of rationality, it is plausible to count this also as an inferior form of free will.

My view has historical roots in philosophy, but if you don't think I'm talking about free will, so be it. I offer a dilemma in reply. In a moment I will characterize an ideal situation implied by Descartes' account of free will. Either the ideal I offer is a characterization of free will or it isn't. If it is, good. If it isn't, then the burden is on those who deny it to show that what is missing is important—to law or otherwise. I argue that the ideal gives us all we could possibly want, so if it leaves out free will, then we don't actually care about free will. So what is the ideal situation? For Descartes, perfect free will would result if two conditions were satisfied: (1) with respect to any statement or course of action, the understanding provides a clear and distinct perception of the truth of the statement or the correctness of the course of action; (2) with respect to any statement or course of action, the will affirms the statement or pursues the course of action if and only if the understanding provides a clear and distinct perception of the truth of the statement or the correctness of the course of action.

Some of the choices we face in life are difficult because, while they can significantly impact whether our lives go well, we are not certain what the result of the choice will be. Which college should I attend? Which career path should I follow? With whom should I pursue a romantic relationship? Faced with such choices, we may lament that we cannot have the wisdom of hindsight *now*. But suppose that for each choice we faced, we had a clear and distinct perception of what the best choice was. Moreover, suppose that our will was constituted so that we always pursued what we clearly and distinctly perceived to be the best choice. The result would be that we would never make the wrong choice, i.e., it would never be the case that an alternative course of action was actually better.

Imagine an ideal Cartesian world—call it "Cartesia"—in which each person satisfies the two conditions of perfect free will. Cartesia undercuts the appeal of incompatibilism, as follows. Either Cartesia is a world in which people have free will or it is a world in which people do not have free will. If it is a world in which people have free will, then free will is compatible with determinism. The compatibility is made clear when we remember Descartes' point that judgments can be free even if the person judging could not have judged otherwise. The crucial thing to see is that even if we accept the incompatibilist's claim that a deterministic world is one in which a person cannot choose or do otherwise, it does not follow that determinism rules out free will. The presumption that a person can choose or do freely only if the person can choose or do otherwise is false, and it should be fairly clear why

Descartes might think this. In the situation he considers, judging otherwise would have led him into error. A world in which he could have judged otherwise thus presents a possibility that he could not rationally want to be actualized, and so nothing important is lost if it turns out that such a possibility is missing from our world.

However, suppose you believe that free will does not exist in Cartesia, i.e., in whatever sense people are free in this world, it isn't the kind picked out by "free will." Perhaps you think that the first, incompatibilist formulation of the power of free choice is closer to what we mean by "free will," and so while it might have been rational for Descartes to judge as he did, insofar as he "could not but judge" the way he did, his judgment was not an exercise of free will. But suppose we modify Cartesia in such a way that, although people still actually satisfy the two defining conditions of this world, such people could have chosen or acted otherwise than they did, in the incompatibilist's sense of "could have chosen or acted otherwise." The challenge faced by those who deny that free will exists in the original Cartesia is to explain why a modified Cartesia in which free will exists is preferable to the original world in which free will does not exist. We can put the challenge as follows. I assume that all parties accept that being rational is preferable to being irrational. Adding free will to Cartesia is tantamount to adding certain possibilities, but given the nature of that world, each possibility we imagine being added is such that its actualization would count as an instance of irrationality. But then saying that a Cartesia in which free will exists is preferable to a Cartesia in which free will does not exist amounts to saying that worlds in which the possibility of irrationality exists are preferable to worlds in which the possibility of irrationality does not exist, despite the fact that being rational is preferable to being irrational.

Given the widely held belief that a necessary condition for moral responsibility is that a person could have chosen otherwise than she did, perhaps we can appeal to the importance of moral responsibility to explain why a Cartesia in which incompatibilist free will exists is preferable to a Cartesia in which incompatibilist free will does not exist. There are two reasons why this appeal is problematic, however. We have already encountered the first reason: the compatibilist will claim that moral responsibility requires compatibilist free will, not incompatibilist free will. The second, and I think more telling, reason has to do with the peculiar nature of Cartesia. In the last chapter I noted that the drama of moral responsibility—the reactive attitudes, and the attendant practices of praising and blaming, rewarding and punishing—unfolds primarily in what I called A-responsibility. It should be evident, however, that this drama would be missing from Cartesia. Because each person in Cartesia believes what she ought to believe and does what she ought to do, the practices of censuring, blaming, and punishing simply will not exist. Moreover, since each person in Cartesia is already D-responsible, no person has a reason to adjust his CAA (since the point of such adjustments is to bring one closer to D-responsibility). Questions about moral responsibility, if they were to arise at all, would not have the urgency that they have in our world. If the importance of free will is parasitic on the importance of moral responsibility, then free will does not seem important in Cartesia. Thus we can agree with Timothy O'Connor when he writes that "an agent who was a 'natural saint,' always and effortlessly choosing the good with no contrary inclination, would not have freedom of will among his virtues."[13]

Earlier I said the presumption that a person can choose freely only if the person can choose otherwise is false, and I used this point to undercut the incompatibilist's claim that determinism rules out free will. But if the presumption is false, then isn't there something wrong with traditional compatibilism as well? Recall that the traditional compatibilist and incompatibilist agree that a person's choice was free only if the person could have chosen otherwise; their disagreement is over how to interpret "could have chosen otherwise." According to the traditional compatibilist, a person in Cartesia could have chosen otherwise than she did if she would have chosen otherwise had relevant circumstances been different. Suppose the relevant circumstances concern the person's clear and distinct perception of something, e.g., the pertinent counterfactual concerns what the person would have willed had she not had a clear and distinct perception. If she would have willed differently, then her willing is free; if she would not have willed differently, then her willing is not free. In Chapter 2 we encountered the charge that the compatibilist reading of "could have chosen otherwise" is irrelevant, and in Cartesia, it *is* irrelevant. As I argued in Chapter 3, the significance of the kind of counterfactuals that the traditional compatibilist reading of "could have chosen otherwise" yields is that knowledge of such counterfactuals can provide us reasons to act in ways that bring us closer to D-responsibility. Because each person in Cartesia is already D-responsible, knowledge of such counterfactuals cannot serve this function in that world; this is why these counterfactuals are irrelevant to Cartesia.

It thus appears that if "could have chosen otherwise" is relevant to free will and moral responsibility but is not relevant to Cartesia, then "could have chosen otherwise" is relevant to free will and moral responsibility only in worlds where people are *not* perfectly rational.[14] Descartes himself anticipates this point. Descartes' chief aim in Meditation Four is to show that human error is compatible with the existence of a perfect God. The putative worry is that, since God is perfect and is the creator of human beings, we would expect that God's creations would be without flaw. Insofar as humans are flawed, it looks like God is flawed as well. Descartes replies to this argument as follows:

> For in truth it is no imperfection in Deity that he has accorded to me the power of giving or withholding my assent from certain things of which he has not put a clear and distinct knowledge in my understanding; but it is doubtless an imperfection in me that I do not use my freedom aright, and readily give my judgment on matters which I only obscurely and confusedly conceive.[15]

The imperfection is that Descartes does not use his freedom aright, but here Descartes claims that it is *his* fault, not God's, that he does not use his freedom aright. But couldn't have God made Descartes such that he always uses his freedom aright? Descartes considers this possibility:

> I perceive, nevertheless, that it was easy for Deity so to have constituted me as that I should never be deceived, although I still remained free and possessed of a limited knowledge, viz., by implanting in my understanding a clear and distinct knowledge of all the objects respecting which I should ever have to deliberate; or simply by so deeply engraving on my memory the resolution to judge of nothing without previously possessing a clear and distinct conception of it, that I should never forget it.[16]

God could have made Descartes in such a way that he never erred, but he didn't make Descartes this way. Isn't Descartes just like Mr. Puppet, then? Descartes flaws seem to be the result of suboptimal design, after all. But Descartes rejects this conclusion:

> I have even good reason to remain satisfied on the ground that, if he has not given me the perfection of being superior to error by the first means I have pointed out above, which depends on a clear and evident knowledge of all the matters regarding which I can deliberate, he has at least left in my power the other means, which is, firmly to retain the resolution never to judge where the truth is not clearly known to me: for, although I am conscious of the weakness of not being able to keep my mind continually fixed on the same thought, I can nevertheless, by attentive and oft-repeated meditation, impress it so strongly on my memory that I shall never fail to recollect it as often as I require it, and I can acquire in this way the habitude of not erring.[17]

Let's note, first, that an appeal to incompatibilist free will isn't doing any work in this response. Descartes' errors result from *not remembering* the rule of affirming only that which is clear and distinct. Imagine a situation in which Descartes forgets the rule and errs by affirming that which is not clear and distinct. If we say that he could have avoided erring in that situation, we are not imagining a situation in which he avoids erring despite forgetting the rule. It is precisely the fact that he forgot the rule that explains why he affirmed that which was not clear and distinct, so we have no reason to believe that he might have honored the rule in the situation in which he in fact forgot it. In saying that he could have avoided erring, we are saying that he would have avoided erring had he remembered the rule. And, of course, it is the truth of this counterfactual that prompts Descartes to attentively meditate on the rule, for the effect of attentively meditating on the rule is that he will remember it in the future and thus avoid further error. If we consider the initial situation where Descartes didn't remember the rule, it seems right to say that his failure to remember was a result of God's design; after all, God could have firmly planted the rule into Descartes' memory, but he didn't. What is interesting is that, instead of denying his responsibility and blaming God, Descartes focuses on the fact that it is within his power to make sure he remembers it in the future.

In the previous chapter I noted that the goodwill demand can play an explanatory role even in those cases in which the goodwill demand is not met. Similarly, the clear and distinct rule—and demands of rationality, generally—can play an explanatory role even in those cases in which the rule is not followed or demands are not met. Descartes did not follow the rule, but insofar as his CAA is structured so as to bring about changes that make it more likely that he will follow the rule in the future, we can say that one function of his CAA is to produce what I will call *D-rationality*. D-rationality is a form of D-responsibility that involves being reliably rational; a D-rational person can be counted on to be rational, as it were. In a related vein, we can characterize *A-rationality* as a form of A-responsibility that involves adapting one's CAA toward greater D-rationality. Just as a moral agent is one for whom the goodwill demand can function as a reason for action, a rational agent in one for whom the clear and distinct rule can function as a reason for action. Whether this rule functions as a reason for action for some agent will depend on whether the agent takes responsibility for her failures to follow the rule, i.e., it will depend on the agent's belief that it was within her power to follow the rule and the fact that she appreciates the normative significance of the rule. In taking responsibility for our lapses from rationality, we can actually make ourselves more rational—and thus more free.

In drawing the parallel between moral agents and rational agents, I do not mean to suggest that moral demands are different in kind from rational demands. On the contrary, legitimate moral demands—as well as legitimate legal demands—are a subclass of rational demands. While this claim may be contentious, any account that aims to preserve the connection between responsibility and rationality is committed to something along these lines. I will come back to this point in Chapter 7; right now I want to say a bit more about the second desideratum, namely, that our account should preserve the connection between responsibility and free will. But first, a note about terminology. In what follows, I will use 'A-responsible' and 'D-responsible' to pick out those aspects of rationality that relate specifically to moral or legal spheres, reserving 'A-rational' and 'D-rational' to characterize rationality more broadly. With this in mind, let's consider what Descartes says about "right" and "wrong" uses of free will:

> But if I abstain from judging of a thing when I do not conceive it with sufficient clearness and distinctness, it is plain that I act rightly, and am not deceived; but if I resolve to deny or affirm, I then do not make a right use of my free will; and if I affirm what is false, it is evident that I am deceived; moreover, even although I judge according to truth, I stumble upon it by chance, and do not therefore escape the imputation of a wrong use of my freedom; for it is a dictate of the natural light, that the knowledge of the understanding ought always to precede the determination of the will. And it is this wrong use of the freedom of the will in which is found the privation that constitutes the form of error.[18]

I can make right use of my free will or I can make wrong use of it, and we've noted that Descartes thinks a person can be responsible for making the wrong use of free will, e.g., affirming the truth of a statement despite lacking a clear and distinct perception of its truth. While we should preserve the idea that a person can be responsible for her errors, Descartes' strict definition of free will complicates this matter. Recall that Descartes characterizes the will in such a way that all acts of will are free. He also says that an action, choice, or judgment results from our will just in case the action, choice, or judgment is not determined by an external force. But if we accept these formulations, it seems like a person can fail to be responsible for her actions, choices, or judgments even if she freely wills them. Consider a person who is on trial for murder. If the person committed the murder but is found to be legally insane, then the person is not legally responsible for the murder. However, saying that the person was legally insane at the time of the murder doesn't carry with it the implication that the murder was the result of an external determining force. It is thus possible that the person is not responsible for the murder even though the person committed the murder freely. Insofar as we want to maintain the connection between free will and responsibility, we cannot accept Descartes' account of free will as it stands. As it turns out, Descartes' notion of "grades of liberty" suggests the amendment we need.

If all acts of will are free, then it is hard to understand how willing in a condition of ignorance is less free than willing in a condition of knowledge. Willing freely is like being pregnant; a pregnant person cannot be more pregnant than another pregnant person. However, insofar as grades of liberty reflect epistemic differences, being more or less free seems to be a matter of being more or less rational. A perfectly free person would thus be a perfectly rational person, and this is precisely the condition of those living in Cartesia. But as we saw, it isn't clear that being perfectly free in the sense of being perfectly rational entails that a person has free will. I argued that if people in Cartesia don't have free will, then free will isn't important, and this suggests that if free will is important in our world, it's importance turns on the fact that we are not perfectly rational, i.e., we are not D-rational. We need an account of free will, then, that implies neither that insane people have free will nor that free will requires D-rationality. An account that characterizes free will in terms of A-rationality does just that. To see this, consider again the person who is found to be legally insane. According to the most widely used legal definition of insanity, an insane person is a person who did not appreciate the criminality of her conduct. If the legally insane lack an important form of knowledge and the legally insane are not responsible for their actions, we may be tempted to conclude that the legally insane are not responsible *because* of their lack of knowledge. But while it has long been recognized that ignorance is sometimes an excusing condition, it does not excuse in all conditions. Whether ignorance is an excusing condition depends on whether the person *should* have known or whether the person should have refrained from acting in her condition of ignorance. To be legally insane, it is not enough that the person lacked knowledge of the criminality of her conduct; the person must also lack the capacity to know the criminality of her conduct. It does not make sense to say that the legally insane person should have known what she was doing because it does not make sense to say that she could have known, and it does not make sense to say that she could have known because, lacking the capacity, she is incapable of acquiring this knowledge. By contrast, an A-rational person can be responsible for her ignorance because recognition of this ignorance provides her reasons for action, reasons that, if effective, tend to bring about greater D-rationality. Free will is threatened, then, in cases where a person is either unable to recognize her ignorance or is incapable of responding to her ignorance in a way that tends to bring her closer to D-rationality.

If D-rationality were simply a function of what one knows, then increases in knowledge would be sufficient to bring one closer to D-rationality. But knowledge is not enough; I must also *will* the correct course of action. The failure to will what I know to be the correct course of action is clearly an instance of irrationality, but am I responsible for such a failure? Once again, the issue turns on whether I am an A-rational agent. Just as an A-rational agent responds to her recognition of ignorance in the direction of greater knowledge, an A-rational agent responds to her failures of will by taking steps to reduce the likelihood of such failures in the future. If my CAA is structured so as to render me incapable of responding in this way, then I am not responsible for the irrationality that results from my failures of will. Maintaining the connection between free will and responsibility thus also requires us to characterize free will as the capacity to respond appropriately to failures of will. Put another way, if having free will is a matter of being

A-rational, and A-rationality is characterized in terms of tending in the direction of D-rationality, then the kind of responsiveness required for A-rationality concerns both understanding and will.

DELIBERATION

In Chapter 2 we encountered the argument that deliberation would be impossible if determinism were true. The argument, recall, is that one deliberates in order to decide what to do, but if what one is going to do has been rendered inevitable by conditions already existing, then one cannot decide whether to perform an action or not. Deciding implies alternatives, and determinism rules out alternatives. However, closer inspection of the phenomenon of deliberation reveals that this argument is flawed. The alternatives required for decision-making are *epistemic*, not nomological.

In the previous section I explained the sense in which the issue of whether a person can choose or do otherwise is insignificant in Cartesia. Deliberation is similarly insignificant in Cartesia, for irrespective of the truth of determinism, no one deliberates in Cartesia. As Taylor notes, "I can deliberate only with the view to deciding what to do, to making up my mind."[19] Deliberation is the reflective process of weighing reasons for and against various courses of action. If one deliberates in order to make up one's mind, then we can say that the point of deliberating is to obtain clear knowledge of what action the balance of reasons favors. There are other processes one might employ to make up one's mind, of course. For example, one could simply flip a coin. Because deliberation involves reflecting on one's reasons for action, deliberation preserves a rational link between our reasons for action and our decisions. However, if it is immediately clear to you which action the balance of reasons favors, reflecting on the reasons for and against various courses of action will have no point. In Cartesia, each person has immediate and clear knowledge of the correct course of action and spontaneously wills what is revealed to be the correct course of action. Doubt, hesitation, and uncertainty are thus absent in Cartesia, as is the need to reflect on reasons for and against various courses of action. The correct course of action will be so obvious to people in Cartesia that no other courses of action will even tempt them, and so the "decisions" people make will look like the only possible "decisions" *to* make. For example, if a person in Cartesia is faced with "deciding" between salad and soup and the balance of reasons favors salad, then at no point will soup appear as a live option. This last claim is potentially misleading, however, so we need to clarify the sense in which an option is possible, or live.

Let's call an option *epistemically possible* for an agent, at a time, just in case at that time it appears to the agent that the option is consistent with the agent's balance of reasons. Suppose you are asked to pick between salad and soup and the balance of reasons favors salad. If the balance of reasons favors salad, then picking soup is inconsistent with the balance of reasons. But although the balance of reasons may favor salad, it may also be true that you do not initially *recognize* this. In failing to recognize this, you fail to recognize that soup is inconsistent with what the balance of reasons favors. Indeed, it is precisely because salad and soup are each epistemically possible that you undertake a process of deliberation, and if all goes well, the epistemic possibilities will be reduced to just salad.

Cartesia is a deterministic world and no one in Cartesia deliberates. As we've seen, however, we do not need to invoke determinism to explain why there is no deliberation in Cartesia. Deliberation is a process we undertake in response to a lack of clear knowledge of what action the balance of reasons favors, and since no person in Cartesia lacks this knowledge, no person in this world deliberates. But when we appreciate this explanation of the absence of deliberation in Cartesia, we can also appreciate why determinism is irrelevant to deliberation. To see this, imagine again a modified version of Cartesia. The modified version is a deterministic world in which a person may initially lack clear knowledge of what action the balance of reasons favors. In such situations, the person initiates a process of reflection that eventually yields clear knowledge. To illustrate the idea, imagine asking a person to pick the correct solution to a multiplication problem, e.g., the correct solution to 1259 x 769. Suppose that the answer options presented are 968, 171 and 956, 283. If the person lives in the original Cartesia, the person will have immediate and clear knowledge that the answer is 968, 171. A person living in the modified Cartesia may not have such immediate and clear knowledge, but this person employs a procedure, e.g., a partial products algorithm, that eventually results in the correct solution. Although 968, 171 is the correct answer, to the person in modified Cartesia this answer will not initially appear as the only epistemically possible solution; after all, if it did so appear, there would be no point in employing the procedure. But though the person initially lacks clear knowledge of the answer, the availability of the procedure insures that her ignorance isn't *permanent*, and the person in modified Cartesia is as free from error as the person in the original.

The person in modified Cartesia undertook the procedure in order to find the correct solution to a multiplication problem. Perhaps making sense of this requires that at one time more than one solution is epistemically possible, but it doesn't require the rejection of determinism. If we think of deliberation as a response to a lack of clear knowledge of what the balance of reasons favors, as we should, then all that is required to make sense of deliberation is that at some time more than one action is epistemically possible. Reflection on ordinary uses of 'deliberation' supports this. Consider a jury that is given the task of delivering a verdict, and suppose it takes the jury three days to reach such a verdict. We would say that the jury *deliberated* for three days—they considered reasons for and against finding the defendant guilty—and the explanation for why the jury took three days is that they did not initially know what verdict the balance of reasons favored and it took them three days to arrive at such knowledge. If the jury had possessed clear knowledge at the outset, there would have been no need to deliberate.

At the outset of deliberations members of the jury may have felt that the verdict could go either way, and if we are thinking about possibilities epistemically, they are correct. However, if determinism is true, these distinct epistemic possibilities do not correspond to distinct nomological possibilities. To think otherwise is to succumb to an illusion, an illusion generated by an initial lack of clear knowledge. But this does not imply that epistemic and nomological possibility are not importantly related. In Cartesia, epistemic and nomological possibility correspond exactly, and this correspondence is explained by the fact that agents in Cartesia are perfectly rational. For any situation an agent in Cartesia encounters there will be just one course of action that is consistent with the agent's balance of reasons, and the agent will have immediate and clear knowledge of what this action is. In being perfectly rational, such agents will only those actions that are consistent with their balance of reasons. Thus in any situation in which there is just one epistemically possible option, there is also just one nomologically possible option. In this way, all inferences from epistemic possibility to nomological possibility are valid in Cartesia. This may suggest that we find nomological possibility wherever we find epistemic possibility. Descartes himself suggests this:

> This indifference, moreover, extends not only to things of which the understanding has no knowledge at all, but in general also to all those which it does not discover with perfect clearness at the moment the will is deliberating upon them; for, however probable the conjectures may be that dispose me to form a judgment in a particular matter, the simple knowledge that these are merely conjectures, and not certain and indubitable reasons, is sufficient to lead me to form one that is directly the opposite.[20]

The line of reasoning supporting this conclusion is fairly straightforward: in cases in which I clearly perceive the truth, my will is so strongly inclined to affirm that which I clearly perceive that I cannot but judge the way I do. By contrast, in cases in which I lack a clear perception of the truth, though my will may be inclined in one direction, the inclination isn't strong enough to guarantee that I actually judge in that direction, i.e., I *can* but judge the way I do. But while distinct epistemic possibilities are *compatible* with distinct nomological possibilities, distinct nomological possibilities do not *follow* from distinct epistemic possibilities. Consider modified Cartesia. There are situations in that world in which more than one option is epistemically possible, but agents respond to these situations by undergoing a process that reduces the epistemic possibilities to one. As I characterized modified Cartesia, the process that agents undergo to reduce epistemic possibilities is always successful, and agents in this world perform only those actions that remain epistemically possible upon completion of the process. Nothing about the agent's response to a lack of clear knowledge—including the process the agent employs to gain clear knowledge, as well as the actions the agent performs in light of the knowledge gained—conflicts with the truth of determinism.

Deliberation is born out of ignorance, but it is important to clarify the kind of ignorance that is—and is not—relevant to deliberation. Taylor correctly states that a person can deliberate only if the person does not already know what they are going to do, but if determinism is true, there are always conditions existing prior to my actions that are sufficient for those actions, and if I can know what those conditions are I can know everything I am going to. In that case, deliberation would be impossible. Taylor goes on to note that this implies only that I cannot deliberate if I know these antecedent conditions. Might I be able to deliberate if such conditions existed but I was ignorant of them? Taylor's reply is that

> It is odd … to suppose that deliberation should be a mere substitute for clear knowledge. Ignorance is a condition of speculation, inference, and guesswork, which have nothing whatever to do with deliberation. A prisoner awaiting execution may not know when he is going to die, and he may even entertain the hope of reprieve, but he cannot deliberate about this. He can only speculate, guess—and wait.[21]

The prisoner may not know when he is going to die because he may not know all the determining factors, but it does not follow that the prisoner can deliberate over when he is going to die. While it is true that a prisoner cannot deliberate about when he is going to die, that is because the prisoner's reflection on what action the balance of reasons favors has no bearing on when he gets executed. By contrast, whether I end up with salad on my plate does depend on my reflection on what menu item the balance of reasons favors. Deliberation is a response to ignorance, but it isn't a response to ignorance of determining factors. It is a response to ignorance of what action the balance of reasons favors. The relevant determining factor involved in my selecting salad is *knowledge* of what the balance of reasons favors, knowledge that is acquired as a result of the process of weighing reasons for and against various courses of action. Of course, one undertakes the process with the assumption that the outcome of this process will have some bearing on what happens, e.g., I cannot deliberate about whether to order salad or soup if it turns out neither is on the menu. But in the typical case whether I end up with salad on my plate is not akin to whether or when the prisoner gets executed. Deliberation is not a luxury afforded by a nomologically permissive universe but a crutch necessitated by an epistemically demanding one. Because deliberation can bring us closer to achieving the ideal of rationality, it's also a crutch that can help us become more free.

At the outset I said that we should aim for an account that explains why questions of legal responsibility turn on questions of rationality, on the one hand, and also explains why questions of responsibility are often framed as questions about free will, on the other. My account does just that. To have free will is to be A-rational, and to be A-rational is to be responsive to rational demands. When I wonder whether a person has free will, I am wondering whether certain demands have normative force on this person. For a demand to have normative force on a person, the person must be able to appreciate the nature of the demand and respond appropriately to failures to meet the demand.

To this point there has been little discussion of the precise content of existing law. The omission is intentional, for if we aim to critique existing law, we need some independent theoretical lens to guide our critique. Two pressing questions for law emerge from the account I have provided. First, if legal responsibility turns on questions of free will, to what extent does existing law adequately address ways in which free will can be compromised? As we will see in Chapter 7, the conception of rationality I offer is more expansive than what is typically reflected in law. If we concern ourselves solely with what the law tends to focus on—knowledge of the criminality of an act—a gap between rationality and responsibility emerges, i.e., the criteria the law uses have people turning out to be rational—and thus responsible—even though there are good grounds for questioning their responsibility. We can close this gap, but only if we adopt the broader conception of rationality provided here. Second, if legal responsibility turns on questions of rationality, to what extent is it rational for people to obey existing law? My account implies that it will make sense to take legal responsibility only if conformance to law is rational. But it may not be, even if we use a fairly permissive sense of rationality. This provides another important complication to the law because if it is not rational to conform behavior to law, then it will not make sense to take responsibility for breaking the law. This issue is addressed in the last chapter.

NOTES

1. Aristotle, *Nicomachean Ethics*, trans. Terence Irwin, in *Readings in Ancient Greek Philosophy*, ed. S. Marc Cohen, Patricia Curd, and C. D. C Reeve (Indianapolis: Hackett, 2005), 868–69.

2. Though I will say a bit more about it in Chapter 7.

3. Joseph Raz, *Practical Reasoning and Norms* (London: Hutchinson & Co., 1975); Michael Smith, *The Moral Problem* (Oxford: Blackwell, 1994); Jonathan Dancy, *Practical Reality* (Oxford: Clarendon Press, 2000). While motivating and explanatory reasons are often treated as equivalent, some have argued that there are subtle differences between motivating and explanatory reasons that enjoin us to keep them distinct (e.g., Kurt Baier, *The Moral Point of View* (Ithaca, NY: Cornell University Press, 1958). For my purposes, however, we can treat them as more or less equivalent.

4. T. M. Scanlon, *What We Owe to Each Other* (Cambridge, MA: Harvard University Press, 1998).

5. Raz, *Practical Reasoning*; Scanlon, *What We Owe*.

6. I'll elaborate on what I mean by "content" in the next chapter.

7. Aristotle, *Ethics*, 869.

8. Davidson, "Actions, Reasons, and Causes."

9. Another issue that arises here is the relation normative reasons bear to *motivation*. *Very* roughly, internalists about normative reasons hold that normative reasons necessarily motivate. Externalists reject this, arguing that a person can have a normative reason to do something and yet not be motivated to do that thing. Though my claims about reasons may look more sympathetic to internalism than externalism, I aim to be neutral on this issue, i.e., I think the account of reasons I offer in this book can be given both an internalist and externalist reading.

10. Rene Descartes, *Meditations on First Philosophy*, trans. John Veitch (1901), accessed July 15, 2016, http://www.wright.edu/~charles.taylor/descartes/meditation4.html.

11. Descartes, *Meditations*.

12. Descartes, *Meditations*.

13. Timothy O'Connor, "Free Will," *Stanford Encyclopedia of Philosophy*, accessed June 3, 2016, http://plato.stanford.edu/entries/freewill/#1.3.

14. In this way, free will is like courage. In worlds where nothing threatens what is good, courage is irrelevant. Our world is not like that, of course, and that is why courage is important.

15. Descartes, *Meditations*.

16. Descartes, *Meditations*.

17. Descartes, *Meditations*.

18. Descartes, *Meditations*.

19. Taylor, "Libertarianism," 405.

20. Descartes, *Meditations*. In this passage Descartes uses "deliberating" in the sense of "considering." This is clear from the context: given what Descartes writes, it is possible to deliberate and possess clear knowledge. However, if one possesses clear knowledge there is no need to deliberate in the sense of weighing reasons for and against various courses of action.

21. Taylor, "Libertarianism," 405.

Chapter 5

Dasein Design

To this point our discussion of the relevance of neuroscience to law has focused on whether the causal determinism to which that science seems committed is compatible with free will and responsibility. As we've seen, the motivating assumption of this discussion is that human beings are purely physical systems—*you are your brain*, as Greene puts it. For Greene, and Eagleman, the philosophically interesting questions concern the *implications* of this assumption. This doesn't suggest that Greene and Eagleman are unaware of the mind-body problem in philosophy; rather, it suggests that they think this problem has been solved, and solved in favor of physicalism. As we saw in Chapter 2, however, there are at least three positions that satisfy the conditions of physicalism—Cartesian materialism, Ryleanism, and functionalism—and how we conceive the relevance of neuroscience to psychological explanation will depend on which position we adopt. In this chapter I argue for a form of functionalism, and my argument largely rests on a consideration of how a great deal of explanation in cognitive science works.

EXPLANATION IN COGNITIVE SCIENCE

A good deal of science is concerned with explaining the properties of complex systems by analyzing such properties into the organized interactions among other simpler systemic properties or properties of component subsystems.[1] This explanation-by-analysis is *functional* analysis because it operates at a level of abstraction that identifies analyzing properties in terms of what they do or contribute, rather than their intrinsic constitutions. As understood here, the targets of such explanations are not individual events, but dispositional properties, which can be specified as regularities. Examples include the photoelectric effect, the Stroop effect, the McGurk effect, Thorndike's law of effect, and Emmert's law. The strategy of explaining the properties of complex systems by functional analysis is ubiquitous in science and engineering, and by no means special to psychology. For example, when we explain how an internal combustion engine works, we describe the interaction of parts specified in terms of the contributions they make to the overall system, e.g., *fuel-injector*, *valve-lifter*, and *air filter*.

Cognitive folk psychology, while often portrayed as a set of laws or truisms, is sometimes portrayed as a functional analysis of intelligent or adaptive behavior. As an illustration, consider the following (toy) analysis of the ability to act so as to satisfy goals, yielding a regularity in behavior that might be called the Planning Effect (Figure 5.1). Though this diagram reveals information about causes, the explanatory targets of this sort of analysis are not particular points in state space or particular trajectories through it. Rather, the aim of this kind of analysis is to appeal to a system's design to explain why one finds the trajectories one does find and not others. The design provides a model of the state space and constrains the possible paths through it, thereby explaining a regularity we find in the system's behavior.[2]

More generally, the explanandum of a functional analysis is a capacity, e.g., the capacity for adaptive behavior, and the strategy is to explain the capacities of a complex system by exhibiting the abstract functional design of that system—to show, in short, that a system with a certain design is bound to have the capacity in question.

Ned Block provides a nice (again toy) illustration of the idea:

> Suppose one wants to explain how we understand language. Part of the system will recognize individual words. This word-recognizer might be composed of three components, one of which has the task of fetching each incoming word, one at a time, and passing it to a second component. The second component includes a dictionary, i.e., a list of all the words in the vocabulary, together with syntactic and semantic information about each word. This second component compares the target word with words in the vocabulary (perhaps executing many such comparisons simultaneously) until it gets a match. When it finds a match, it sends a signal to a third component whose job it is to retrieve the syntactic and semantic information stored in the dictionary. This speculation about how a model of language understanding works is supposed to illustrate how a cognitive competence can be explained by appeal to simpler cognitive competences, in this case, the simple mechanical operations of fetching and matching.[3]

The target of this explanation is a capacity, and the explanation works by specifying a functional design, i.e., the way components—specified in terms of the *jobs* they perform—are organized. In specifying a functional design, i.e., in specifying the way components—described in terms of the jobs they

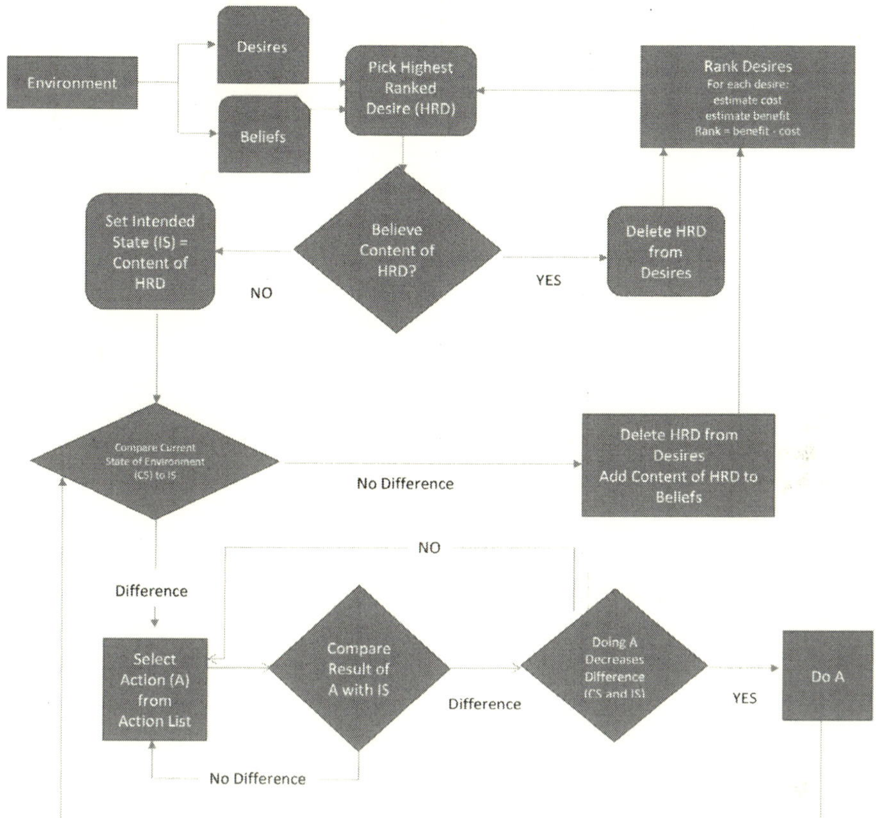

Figure 5.1 A toy analysis of the Planning Effect. Author created.

perform—are organized, some causal consequences of components are highlighted and others are ignored. In the context of a successful design explanation, the causal consequences highlighted by functional specifications are precisely those consequences that enable us to understand why a system with a particular organization of components has the capacity we are trying to explain. In this way, functional specifications provide *causal relevance filters*: by selecting from the myriad causal consequences of a system's components those that are relevant to the target capacity, functional specification makes it transparent why anything that possesses a certain organization of components is bound to have the capacity in question. Without this filtering, we are simply left with a welter of noisy detail with no indication of what is relevant and what is a mere by-product of this or that implementation. Furthermore, by abstracting from the implementation details that are irrelevant to the achievement of the target capacity, causal relevance filtering enables us to understand why systems that differ in implementation details but retain the overall design all exhibit the target capacity.

Functional analysis is ubiquitous in cognitive science, but explanation by functional analysis is found throughout the sciences. So what, if anything, is distinctive to the functional analyses we find in cognitive science? In Chapter 2 I noted how work in computing theory led many philosophers and scientists to take seriously the hypothesis that the mind is a computer, a hypothesis that would turn the science of mind into "a science of structure and function divorced from material substance." A widely influential articulation of this vision was provided by the cognitive scientist David Marr. According to Marr, cognitive science is concerned with three basic projects.[4] The first project involves specifying the task we are trying to explain, e.g., a capacity. The second project is to specify the way a system represents various aspects of the task domain and the procedures the system employs for accomplishing the task. The third project is to show how these representations and procedures are implemented in a physical system. Suppose we want to know how a pocket calculator works. To answer this question, we first need to say more precisely what it is about the calculator we are trying to explain. For example, we might say that we want to know how the calculator is able to multiply. This would correspond to the project of specifying the task. The next project would be to explain this task by specifying the way the calculator represents numbers and specifying the kinds of operations the calculator performs on those representations. Think about when you learned to multiply. You were taught an algorithm (a procedure) for manipulating numerals (representations of numbers). The representations and procedure are such that, given any well-posed multiplication problem, you can apply the procedure to the numerals to get a correct answer.[5] That is how we explain your capacity to multiply. We explain the calculator's capacity to multiply in the same way, e.g., by specifying how it represents numbers and by specifying the procedure it uses to manipulate those representations. The last step is to show how those representations and procedures are physically embodied in the calculator. To complete this step, we need to know something about the calculator's electronics. Crucial to this division of labor is the idea that simply knowing the electronics isn't enough to understand how the calculator multiplies. We also need the description given in terms of representations and procedures, a description that is more abstract than what is provided by a description of the calculator's electronics.[6] In short, we need the causal relevance filters that functional characterization provides. What makes cognitive science special is the particular nature of these filters, and if there is anything distinctive about the functional analyses we find in cognitive science, it is the fact that the causally relevant contributions are contributions made by representations, or states with *content*.

REPRESENTATION IN COGNITIVE SCIENCE

A pervasive assumption in cognitive science is that the mind/brain employs and processes representations in the service of reasoning, thinking, categorizing, planning, problem-solving, understanding, imagining, remembering, and perceiving.[7] This assumption can be challenged, however, and most cognitive scientists agree that appeals to representation earn their place in the science by providing the best explanation of the aforementioned phenomena. As noted, modern scientific theories of cognition—and perception, I should add—are largely inspired by the computer model of the mind, or what Alan Newell and Herbert Simon dubbed the "physical symbol system hypothesis."[8] According to this hypothesis, cognition consists in manipulating inner symbols (e.g., data structures) according to rules (e.g., programs). This view is perhaps most famously associated with the work of Jerry Fodor, who argues that the human mind/brain employs a language-like scheme of representation (the language of thought [LOT]) and that cognitive processes are computational processes defined over the formal properties of these representations.[9] Such accounts of the architecture of human cognition are often called *classical*, and are typically contrasted with various *connectionist* or *neural network* architectures. However, insofar as both classicists and connectionists attempt to explain cognitive and perceptual capacities in terms of representations, both are committed to some version of *the representational theory of mind* (RTM).

RTM has been discussed, defended, and challenged since the time of Aristotle, and philosophers interested in RTM have been principally preoccupied with trying to give an account of the nature and possibility of representation. Specifically, they have been preoccupied with trying to make sense of *intentionality*, or the *aboutness* of representations. For example, consider a picture hanging on a brick wall. Physically speaking, there appears to be nothing very special about either the painting or the wall. And yet, whereas we do not typically think that brick walls are *about* anything, somehow pictures are: they are *about* sunsets and graduations and dogs playing poker. How is this possible? While it may be tempting to say that pictures get their aboutness from our thoughts, this only pushes the problem back a step. Most contemporary philosophers of mind accept that, in one way or another, mere arrangements of matter are sufficient for the intentionality of the mental. Moreover, the dominant tradition in cognitive science hopes to appeal to representations to explain how humans and other creatures manage to survive and reproduce in a complex and often hostile world, and it is widely agreed that an account of mental representation has to meet certain constraints if appeals to mental representations are to have a legitimate place in science. One constraint is that a theory of mental representation should not presuppose anything cognitive. Because RTM seeks to explain cognitive phenomena in terms of representation, it would be obviously regressive to turn around and account for representation in cognitive terms. Another constraint is that a theory of mental representation should be consistent with what we know (or will come to know) about the brain. Finally, and perhaps most obviously, a theory of mental representation should be consistent with the actual explanatory roles representations are supposed to play in cognitive science. In light of these constraints, what shape might a viable theory of mental representation take?

The constraints put on by the science are typically thought to apply to theories of mental representation in at least two ways. First, they apply to what we can call the *vehicles* of representation. When we understand mental representations as vehicles, we may be thinking of representations either as elements of a scheme of representation (e.g., words or sentences of a language, vectors in a multidimensional activation space) or concrete tokens (e.g., marks on a chalkboard, neural activity in the prefrontal cortex). Second, they apply to how representations (understood as vehicles) come to represent in the first place, or how they get their *content.* Content is often characterized as what a representation is *about*, or *means*, or *stands for*, or *refers to*, or *denotes*, or *expresses*, where the contents of mental representations might include objects, properties, structures, propositions, states of affairs, and perhaps more. The assumption that content is explanatorily relevant is appealing to many, and much effort has been spent trying to find room for content in a methodologically sound science of the mind. Part of the reason for this is that the notion of representation is taken as an explanatory *primitive* in many areas of psychology, and so the lack of a viable account of what it is for one thing to represent something else would threaten much of psychology with bankruptcy.[10] Philosophy can assist here by elucidating a conceptual framework for thinking about representation and content that enables a fruitful way to express and integrate the aims, results, and methodology of the science. However, if we approach philosophical issues about representation and content with an eye toward the science, then we immediately face the problem of showing that a theory of content is compatible with the roles mental representations are deemed to play in our cognitive lives. What are these roles? As John Haugeland points out, in order to survive, sophisticated creatures need to adjust their behavior to features or structures in the environment that are not always present to the creature.[11] For example, successful planning requires that an organism take into account future events, which are by definition not currently present to the organism. Following Andy Clark, we can think of the capacity to plan as "representation-hungry," since it is with respect to explaining capacities like planning that appeals to representation seem to make a good deal of sense.[12] For example, if a creature harbors something that functions as a stand-in, proxy, or surrogate for temporally and spatially distant states of affairs, then the creature could reason about those states of affairs by reasoning about the stand-in.[13]

With the above kinds of examples in mind, some philosophers suggest that mental representations can serve as effective surrogates because mental representations share structure with the domains they are intended to represent. The thought here is that if representational content is grounded in shared structure, then creatures can cognize the *world* by structurally transforming their representations.[14] According to these "picture" theories of content, representing is a relation between structures: a structure R represents a structure C when there is a relation preserving mapping of the objects in R into the objects in C. A consequence of this view is that representations are not symbols that merely refer to or denote objects and properties. Gabriele Contessa provides a nice illustration of the point.[15] The official logo of the London Underground is a red circle with the word 'underground' written across it. This logo denotes the London Underground, and in that sense has the London Underground as its content. But the logo does not represent the London Underground in the same way that a map of the Underground does, for the latter can be used as a surrogate in successful reasoning about how to get from one place to another, about how far a station is from your destination, and so on. Furthermore, the map allows such surrogative reasoning precisely because the map, unlike the logo, shares structure with the line system of the London Underground. As the familiar example of language makes clear, denoting or referring does not require that symbols share structure with the things they denote; when it comes to referring or denoting, the structure of symbols is arbitrary. Thus, any correlation found between the structure of the symbols and the structure of the things the symbols denote would seem purely accidental. Of course, there is nothing wrong with calling symbols that refer or denote without sharing structure with the things they denote "representations," and as will be discussed below, there are philosophically powerful reasons for doing so. However, insofar as much of the explanatory payoff of representations derives from their use as surrogates in such tasks as planning, and insofar as effective surrogacy requires that structural transformations on representations track transformations in the domain represented, there may be good theoretical reasons to reserve the term 'representation' for those vehicles whose structure is relevant to their explanatory role.

Whatever merits there may be in restricting the use of 'representation,' encouragement for thinking of representation in terms of denotation or reference can be found in the ways cognitive science was initially characterized. Recall that the modern version of RTM has its roots in the physical symbol system hypothesis, and since the most familiar example of a symbol system is language, it was perhaps inevitable that work in the philosophy of language would inform a theory of mental representation. One route to this idea begins with the claim that the LOT hypothesis best explains the productivity of thought, or the ability to entertain an indefinitely large number of semantically distinct thoughts.[16] Since productivity seems to require that LOT have a compositional semantics (the meaning of a complex expression is a function of the meaning of its parts and the way in which those parts are combined), and since it is widely assumed that the best account of semantic compositionality for languages is given in terms of references, extensions, and satisfaction conditions,[17] it is perhaps not surprising that representation is often understood in terms of reference. In order for this program to succeed, however, an account of how references and extensions of primitive LOT symbols are fixed is needed. Taking a cue from H. P Grice's observation that there is a sense of 'means' that is roughly equivalent to 'carries information' (e.g., "those clouds mean rain"), a major research goal over the last 30 years has been to articulate and defend some version of the idea that *meaning is information*.[18] Though there are different flavors of information theories of content, most of them share a commitment to the idea that the content of a representation R is in some way determined by what causes tokenings of R. According to this story, for example, the LOT symbol type CAT represents the property of being a cat in virtue of cats causing tokenings of CAT.[19]

One major appeal of causal theories is that they look poised to provide a plausible account of the content of perceptual states. If we think of perceiving as a kind of detecting or indicating, it makes sense to identify the content of perceptual states with the properties that are causally responsible for their tokening. Calling these instances of representation, however, obscures the distinction between detecting/indicating and those cases where appeals to representation get much of their explanatory power. For instance, since CAT has no semantically significant structure, simply harboring the symbol CAT will not allow a creature to use that symbol as a surrogate in reasoning about cats.[20]

The points just considered suggest that there may be good *theoretical* reasons not to call everything that has content a representation. As previously mentioned, the tendency to equate having content with being a representation is encouraged by the ways representation and content are often characterized. For example, if we think of representations as anything having content, proceed to characterize content as whatever something is "about" or "of" or "stands-in for" or "means" or "denotes" or "refers to," and then analyze meaning or denoting in terms of indicating, then indicating inevitably counts as representing. Again, there is nothing wrong with characterizing content in all these ways, as long as we recognize that not all of these ways of characterizing content may be ways of characterizing representation. Unfortunately, these ways of characterizing representation and content invite a *further* conflation of two notions that should be kept distinct: content and target.[21]

To motivate the need for the distinction, and to show why the distinction is often overlooked, notice that a schema like "x is a picture of y" is ambiguous. To see this, consider the following case. A store has been robbed, and a woman gets a clear look at a man whom she suspects committed the robbery. The police ask the woman to give the police artist a description of the man, but for whatever reason the artist's rendering looks nothing like the man witnessed. The police distribute the picture anyway, and two hours later a person is picked up who everyone is prepared to say strongly resembles the artist's depiction. However, the man picked up looks nothing like the man the woman witnessed. How are we to make sense of this? What is the picture *of*? A reasonable response is to say that there are two ways to understand "of" in expressions like "x is a picture of y." The picture is "of" the man the woman witnessed, in the sense that this man is the *target* of the artist's drawing—it is he whom the artist is trying to depict. However, the picture is a picture "of" the man picked up, in the sense that this man *satisfies* the picture—the picture accurately portrays *him*. To avoid confusion, we can call whatever in fact satisfies a representation its "content," reserving the term "target" for whatever it is that is supposed to satisfy the representation.

It is not too much of a stretch to say that what overall shape you think a theory of mental representation will take depends heavily on your view of the propositional attitudes—the beliefs, desires, and intentions that are central to our "common sense" view of the mind. One often finds that discussions of representation and content begin with the propositional attitudes, with the goal being to develop a naturalistically acceptable version of RTM that can accommodate them. In fact, much of the philosophical literature on representation and content gives the impression that the only job of RTM is to give an account of the propositional attitudes. Unsurprisingly, then, skepticism about the propositional attitudes is often interpreted as skepticism about RTM, which in turn is treated as skepticism about representationalist cognitive psychology.[22] As we have seen, there is some merit in this treatment: the dominant view in recent philosophy has been to think about representation in terms of truth or satisfaction conditions, with beliefs being the paradigm case of states with truth conditions. If one abandons the propositional attitudes, then there needs to be some other way to understand what representation is if we are going to salvage RTM. To many, LOT and the truth-conditional semantics that comes with it look like the only game in town.

However, alternative accounts of representational content have emerged over the past decade, accounts that seek to replace reference and truth as the fundamental semantic notions. Drawing on work in computational neuroscience, Paul Churchland has developed an account of representation which he calls "state-space semantics" (SSS).[23] According to SSS, the locus of representational content is the partitioned activation space of a neural network. In neural networks, the spread of activity from one layer of units to another layer of units is mediated by the connection strengths, or weights, between the layers. Learning in such networks consists in adjusting the weights between layers of units, and during learning, the weights come to partition the activation space of a layer of units into salient points or regions. These points or regions are "centers of gravity" around which activity tends to cluster, and the structure of the partitioned space can be described in terms of the geometric relations that obtain between these points/regions. According to Churchland, these activation spaces are properly called *representations* because their geometric structure mirrors the structure of an aspect of the world being cognized, and it is in virtue of these spaces exhibiting the structure of various worldly domains that the network successfully categorizes, navigates, and accomplishes other tasks. In effect, these spaces function as maps that automatically steer a network's activity. Moreover, Churchland's account of representation is *internalist*, in the sense that the representational content of partitioned activation spaces is determined by the geometric relations that obtain between the points and regions of the space, and not whatever causal relations that may exist between mind and world. Additionally, these activation spaces are not properly thought of as true or false of their targets, but more or less accurate along several and simultaneous dimensions. Thus, whereas reference and truth play a central role in LOT, SSS seeks to abandon those notions as the key to understanding the representing activity of brains.

While many of the fine points raised in this discussion of representation and content are somewhat tangential to the overall argument of the book, their inclusion underscores the central explanatory role that representation and content play in cognitive science. Given this central role, it is not implausible to think that representation and content will factor heavily in any attempt to fuse the manifest and scientific images. I return to this point in the next chapter.

NONREDUCTIVE PHYSICALISM AND LEIBNIZ' GAP

There is a long standing issue in the philosophy of science over whether psychology reduces to a more "fundamental" science, e.g., neuroscience. The issue of inter-theoretic reduction generally is largely driven by what I will call the *nomic conception* of science. The nomic conception has two easily recognizable and internally related components: (1) *The content of a science is captured by its laws.* The various sciences are individuated by their theories, and a theory is a collection of laws expressed in a proprietary vocabulary. Thus, ultimately, the content of a science is captured by a set of laws;[24] (2) *Scientific explanation is subsumption under law.* According to the deductive-nomological (D-N) account of explanation,[25] the explananda of science are events and laws. Explanation of an event is accomplished by deducing its occurrence from laws and initial conditions. Explanation of a law is accomplished by deducing it (or a close approximation) from more basic laws and boundary conditions. From the perspective of the nomic conception, the sciences form a reduction hierarchy to the extent that the laws of one science can be deduced from the laws of another science. For example, if the laws of chemistry can be deduced from the laws of physics, then chemistry reduces to physics, and physics is "below" chemistry in the hierarchy. Reduction is just D-N applied between sciences, and for those in the grip of the nomic conception, the goal of unifying science became the goal of providing between-science deductions.[26]

A notable feature of D-N is that, given its deductive structure, the explanations it generates are transitive. For example, if there are laws L_1 and L_2 such that L_2 subsumes L_1, and if L_1 subsumes event E, then L_2 subsumes event E. If we accept the account of reduction sketched above, it follows that if there is some event that is subsumed by the laws of psychology but is not subsumed by the laws of neuroscience, then psychology does not reduce to neuroscience. When we put the matter this way, it should be clear why the conventional wisdom is that functionalism in psychology blocks the reduction of psychology to neuroscience.[27] According to functionalism, psychology discloses nomic connections between functional states. If functional states are multiply realized (e.g., if some instantiations of functional states are not instantiations of biological states), then the laws of psychology subsume events that are not subsumed by the laws of neuroscience, and so the reduction of psychology to neuroscience fails. Similarly, if nomically distinct biological states can realize the same psychological state, then, once again, the laws of psychology will fail to reduce to the laws of neuroscience. Psychology, in this case, will be autonomous from neuroscience.

If we take seriously the idea that laws of psychology subsume events not subsumed by laws of neuroscience, then there is something wrong with the hierarchy that the nomic conception presumes. The problem isn't merely that the hierarchy fails to be a reduction hierarchy, for that hierarchy is committed to the idea that if the laws of a science subsume some event, then laws of the sciences below also subsume that event. The aforementioned considerations of multiple realization purport to show that laws of psychology do not stand in this relation to laws of biology, however, so the argument from multiple realization would require us to abandon the idea that biology is "below" psychology. Nonetheless, accepting the argument from multiple realization does not require that we abandon the spirit of the hierarchy. Reductions require type-type identities, but functionalists can opt for the weaker claim that each token of a psychological state type is identical to a token of some physical state type or other. This would preserve the idea that any event subsumed by the laws of psychology is subsumed by some physical law or other, so failure of reduction would be compatible with the claim that physics is more basic than psychology.[28]

However, if the explanatory contribution of an unreduced, "higher-level" law is simply the set of events it subsumes, it is not obvious that embracing token-token identities (without type-type identities) helps the defender of autonomy. After all, if physical laws subsume any event that unreduced laws of psychology subsume, it seems to follow that physical laws can explain any event explained by unreduced laws of psychology. But if that is so, then unreduced psychological laws are gratuitous. Thus the challenge to the defender of autonomy is to show that nonreduced, higher-level laws come with explanatory benefits not enjoyed by lower-level laws. Of course, defenders of autonomy do think that higher-level laws enjoy such benefits: the explanations provided by higher-level laws are more *general* than the explanations provided by lower-level laws. Even if generality is not the only explanatory virtue, the defender of autonomy seems to be on firm ground in claiming that it is a virtue reserved for explanations that invoke higher-level laws.

Following Elliot Sober, we can think of the choice between higher- and lower-level explanations as involving a trade-off between *breadth and depth*. Here is how Sober puts the point:

> The goal of finding "relevant" laws cuts both ways. Macro-generalizations may be laws, but there also may be laws that relate micro-realizations to each other, and laws that relate micro-to macro-as well. Although "if P then Q" is more general than "if Ai then Bi," the virtue of the micro-generalization is that it provides more details. Science aims for depth as well as breadth. … Returning to Putnam's example, let us imagine that we face two peg-plus-board systems of the type that he describes. If we opt for the macro-explanation of why, in each case, the peg goes through one hole but not the other, we will have provided a unified explanation. We will have explained similar effects by describing similar causes. However, if we choose a micro-explanation, it is almost inevitable that we will describe the two systems as being physically different, and thus our explanation will be disunified. We will have explained the similar effects by tracing them back to different types of cause. Putnam uses the terms "general" and "invariant" to extol the advantages of macro-explanation, but he might just as well have used the term "unified" instead.[29]

Here we find an interesting twist: while multiple realization blocks the kind of unity that between-science deductions were supposed to provide (reduction), it turns out that higher-level laws provide a kind of unity—unity as generality—that depends on multiple realization. On this view, the explanatory virtue of higher-level laws depends on *actual* multiple realization, as opposed to merely possible multiple realization. If the states picked out by higher-level laws are not multiply realized, then the generality provided by higher-level laws would be no greater than the generality provided by lower-level laws. And because explanations given in terms of lower-level laws provide depth (in Sober's sense, i.e., provide micro-details), lower-level explanations would always emerge as superior: equivalent breadth and more depth. This is why reductive explanations are so appealing in the first place.

While the above story about reduction, multiple realization, and autonomy may be intuitively compelling, the autonomy of functional explanation does not depend on actual multiple realization, and while functional explanations may in fact exhibit the kind of generality that is thought to apply to higher-level laws, the primary virtue of functional explanation is not its generality. I will explain why in a moment, but first consider the following passage from Hilary Putnam, as well as Sober's response to it. Putnam writes, "Even if it were not physically possible to realize human psychology in a creature made of anything but the usual protoplasm, DNA, etc., it would still not be correct to say that psychological states are identical with their physical realizations ... such an identification has no explanatory value in psychology."[30] Sober calls Putnam's remark "curious" and notes the following:

> If we take Putnam's remark seriously, we must conclude that he thinks that the virtue of higher-level explanations does not reside in their greater generality. If a higher-level predicate (P) has just one possible physical realization (A), then P and A apply to exactly the same objects. Putnam presumably would say that citing A in an explanation provides extraneous information, whereas citing P does not. It is unclear how this concept of explanatory relevance might be explicated.[31]

Viewed through the lens of the nomic conception, Putnam's remark *is* curious. Because the nomic conception must identify any explanatory bonus yielded by psychological laws with breadth, it weds the autonomy of psychology to actual multiple realization. Yet Putnam appears to be claiming that psychological explanations would be autonomous even if psychological states were not actually multiply realized. Putnam is right, but the nomic conception makes it impossible to see *why* he is right.

Recall that functional analysis is the strategy of explaining the dispositional properties of a complex system by exhibiting the abstract functional design of that system—to show, in short, that a system with a certain design is bound to have the (typically dispositional) property in question. Designs can do this because functional terms pick out the causal powers that are relevant to the capacity being analyzed, i.e., functional terms are causal relevance filters. Moreover, it is precisely the transparency provided by this filtering that enables us to understand *why* anything that possesses these states, processes, or mechanisms is bound to have the target capacity in question. Causal relevance filtering is, therefore, just abstraction from the implementation details that are irrelevant to the achievement of the target capacity. In this way, the possibility of multiple realization is an inevitable consequence of causal relevance filtering, and so it should come as no surprise to find that functional analyses subsume causal paths that have heterogeneous implementations.

It would, however, be a mistake to wed the explanatory power of functional analysis to assumptions about actual multiple realization, for even if there is only one nomologically possible way to implement a design, giving implementation details that go beyond what is specified by an analysis adds nothing to the explanation provided by the design. For example, suppose there is just one nomologically possible way to implement a doorstop—say, by being a particular configuration of rubber. In this case, it would be plausible to hold that being a doorstop—the type—is identical to being a particular configuration of rubber—the type. Because type-type identities give you property reductions, being a doorstop would thus reduce to being a particular configuration of rubber. But a functional analysis that specifies something *as a doorstop* would still be autonomous, in the following sense. Being a particular configuration of rubber comes with any number of causal powers. One of those powers is stopping doors, and in the context of the imagined functional analysis, stopping doors is the only causal power of this particular configuration of rubber that matters to having the target capacity. If we replace 'doorstop' with 'rubber configured thus and so' in our analysis, we won't lose anything as far as the causation goes. However, we will lose the transparency functional analysis affords *unless we specify explicitly that stopping doors is the relevant causal power.* But then the explanation is tantamount to the explanation given in terms of 'doorstop,' i.e., the explanation does not give us anything beyond what is provided by the functional analysis itself.

If we focus on the causal explanation of events and assume type-type identity, then framing explanations in terms of 'doorstop' is guaranteed to give you nothing beyond what framing explanations in terms of 'rubber configured thus and so' gives you, and this is why it has been generally assumed that reduction is incompatible with autonomy. From the perspective of functional analysis, by contrast, autonomy can live with reduction. Design explanations are autonomous in the sense that they do not require "completion" by annexing implementation details, e.g., in the case imagined above, it is irrelevant to explaining the target capacity whether a specific doorstop is a particular configuration of rubber. But design explanations are also autonomous in the sense that adding implementation details would *undermine* the transparency provided by causal relevance filtering and thereby obviate the understanding provided by the design. A doorstop may *be* a particular configuration of rubber, but replacing 'doorstop' with 'rubber configured thus and so' masks the information needed to understand why a system has the target capacity. This is the lesson of Putnam's passage, but the lesson is lost if we try to understand it as a lesson about causal explanations of events.

I am sympathetic to the thought that complete knowledge of implementation details would contribute to a fuller understanding of those systems whose capacities are targeted by functional analysis. Indeed, such details are necessary for understanding how a system manages to have the very causal powers that are picked out by functional analysis. But having a fuller understanding of a system, in this sense, is not the same thing as having a more complete explanation of the capacities targeted for functional analysis. For example, when we analyze the capacity to multiply numbers in terms of a partial products algorithm, the specification of the algorithm tells us nothing about the states, processes, or mechanisms of a system that implements the algorithm (except in the trivial sense that the states, processes, or mechanisms of any system that implements the algorithm are sufficient for implementing it). However, as far as explaining the capacity goes—what we might call the "multiplication effect"—the analysis provided by the algorithm is complete, i.e., the analysis allows us to understand why any system that has the capacity for computing the algorithm *ipso facto* has the capacity to multiply. Because details about how the algorithm is implemented add nothing to the analysis, such details are irrelevant to the explanation of the capacity. If you claim that the presence of a doorstop explains the fact that the door is open, you need to find some doorstop or other interacting with the door. Having found this, however, the fact that it is rubber rather than wood adds nothing to the explanation itself.

The perspective I have outlined here suggests that we replace Sober's distinction between breadth and depth—a distinction that really only makes sense within the nomic conception—with a distinction between *horizontal* and *vertical* explanation. Horizontal explanations answer the question "Why does the system possess capacity C?" by specifying an abstract functional design D. Vertical explanations answer the question "How is design D realized in the system?" by specifying mechanisms and/or processes that perform the relevant functions. Horizontal explanations tell us how a system works, but you can understand how a system works—understand why a system with a certain design has capacity C—without knowing how the design is implemented.

Acknowledging the distinction between horizontal and vertical explanations amounts to acknowledging some form of Leibniz's gap, but pace Leibniz, this does not require embracing dualism. Or rather, while we might need to accept some form of *explanatory* dualism, this does not require us to accept *metaphysical* dualism. To understand how the brain generates mind, we need to know how the brain processes information. But knowledge of physiology does not, by itself, reveal how the brain processes information, and so knowledge of physiology does not, by itself, provide us with the kind of understanding we are looking for. However, achieving the kind of understanding we are looking for does not require us to posit nonphysical stuff; rather, it requires the kind of causal relevance filtering provided by functional analysis, i.e., it requires us to say *which* properties of the brain matter to mind.

As mentioned earlier, embracing functionalism lead many to embrace the autonomy of functional analysis. In cognitive science, the basic idea of the autonomy thesis is that explaining a psychological capacity does not require knowledge of neural mechanisms. While this seems like a straightforward consequence of abstracting from implementation detail, some argue that this kind of abstraction actually works *against* the claim of autonomy. For example, according to Gualtiero Piccinini and Carl Craver,

> Autonomist psychology—the search for functional analysis without direct constraints from neural structures—usually goes hand in hand with the assumption that each psychological capacity has a unique functional decomposition (which in turn may have multiple realizers). But there is evidence that … several functional decompositions may all be correct across different species, different members of the same species, and even different time-slices of an individual organism. Yet the typical outcome of autonomist psychology is a single functional analysis of a given capacity. Even assuming for the sake of the argument that autonomist psychology stumbles on one among the correct functional analyses, autonomist psychology is bound to miss the other functional analyses that are also correct. The way around this problem is to let functional analysis be constrained by neural structures—that is, to abandon autonomist psychology in favor of integrating psychology and neuroscience.[32]

This argument clearly conflates explanatory autonomy with *confirmational* autonomy. If a capacity admits of more than one analysis, merely providing *an* analysis will, of course, leave open the question of whether the analysis provided correctly describes how a system manages to have the capacity in question (assuming it does have the capacity). Knowledge of neural structures is undoubtedly relevant to settling the question of which analysis is correct, but bringing such knowledge to bear in this instance would be an exercise in confirming a proposed analysis, not explaining a capacity. Suppose there are two possible analyses, A and B, for some capacity C, and the neurological data suggest that analysis A is implemented in system S. The explanation of capacity C in S is provided *by A*, not by the neural structures evidence about which confirms A.[33]

Arguably, nothing enjoys confirmational autonomy from anything else.[34] As such, neuroscience that makes well-confirmed psychological capacities impossible or unlikely needs revision as much as a design hypothesis in psychology that appears to have no plausible neural implementation. We can thus agree with Piccinini and Craver's claim that "psychologists ought to let knowledge of neural mechanisms constrain their hypotheses just like neuroscientists ought to let knowledge of psychological functions constrain theirs."[35] This is not an argument against the explanatory autonomy of psychology; it is rather a consequence of it.

Defending the autonomy of functional analysis is not the same thing as defending the autonomy of psychology. Functional analysis is an explanatory strategy, not a scientific discipline, and when we are careful to distinguish horizontal and vertical explanations, and distinguish confirmation and explanation, the autonomy of functional analysis emerges as unproblematic. It also emerges as ubiquitous, for the function-implementation relation iterates: what is function at one level of abstraction is implementation when viewed from a higher level of abstraction, and what is implementation at one level of abstraction is function viewed from a lower level of abstraction.[36] Thus, a resistor is anything in a circuit across which there is a drop in potential. This might be a semiconductor, or a motor or a light bulb: each of these might implement the resistance function. But we have just identified the implementations *themselves* in familiar functional terms, and there are many ways to implement semiconductors, light bulbs, and motors. It is thus no wonder that functional analysis is ubiquitous in the sciences, including neuroscience.

I want to draw three conclusions from this. First, because the function-implementation relation iterates, we can think of the relation as generating (a potentially large number of) function-implementation hierarchies. For example, with respect to any functional analysis, there may be a number of branching paths "down" the hierarchy. So, too, there may be a number of branching paths "up" the hierarchy. Imagine a tangled graph of function-implementation hierarchies, crisscrossing in various ways. Second, we find autonomous explanations wherever we find a functional analysis of a capacity, so we can expect to find autonomous explanations up and down function-implementation hierarchies. But to repeat: such autonomy should not be confused with *confirmational* autonomy. Nor should it be confused with the claim that functional analyses cannot themselves be explained. They can, and this is precisely what vertical explanations do. But this does not undermine the autonomy of functional analysis, for the explananda of vertical explanations are designs—*not* the capacities designs explain. Third, when we travel "up" or "down" function-implementation hierarchies, we do not find principled borders between the sciences. Patricia Churchland and Terrence Sejnowski provide a nice expression of this point:

> The idea that there is essentially one single implementational level is an oversimplification. Depending on the fineness of grain, research techniques reveal structural organization at many strata: the biochemical level; then the levels

> of the membrane, the single cell, and the circuit; and perhaps yet other levels such as brain subsystems, brain systems, brain maps, and the whole nervous system. But notice that at each structurally specified stratum we can raise the functional question: What does it contribute to the wider, functional business of the brain?[37]

While this passage may suggest that the function-implementation hierarchy described rests squarely within neuroscience, Churchland and Sejnowski go on to show why this isn't the case: "without the constraints from psychology, ethology, and linguistics to specify more exactly the parameters of the large-scale capacities of nervous systems, our conception of the functions for which we need explanation will be so wooly and tangled as to effectively smother progress."[38]

Just so. When we look at any given research program that seeks to understand how a natural system works, we find the work spreading out among disciplines like a slime mold. Psychological and neurological capacities exist, but these are not distinctions of kind that have any serious significance for psychology and neuroscience. There is just a lot of analysis to be done to explain capacities, along with the attendant job of discovering how designs are implemented. If we try to draw disciplinary lines on the function-implementation graph—east-west or north-south—we are not going to get anything like a straight line, and anything we do get will have little import for the business of science as such. We have to consult guides that work in other buildings to travel far either north-south or east-west in our quest for understanding. This is division of labor and interest and method, not of subject matter per se.

Functional analysis is ubiquitous in the sciences at every level. If explanation by functional analysis undermines reduction, then reduction is undermined everywhere, in physics and chemistry as well as in economics, psychology, and biology. In this respect, then, there is nothing special about the mind: function-analytical explanations exhibit explanatory autonomy wherever they are found, and they are found everywhere in science, engineering, and everyday life. While the distinction between horizontal and vertical explanations does imply explanatory dualism, it does not imply metaphysical dualism. As we'll see in the next chapter, this explanatory dualism is the key to understanding how the manifest and scientific images can be fused.

NOTES

1. Robert Cummins, *The Nature of Psychological Explanation* (Cambridge, MA: MIT Press, 1983).

2. This idea is developed further in Martin Roth and Robert Cummins, "Two Tales of Functional Explanation," *Philosophical Psychology* 27 (6) (2014): 773–788.

3. Ned Block, "The Mind as the Software of the Brain," in *An Invitation to Cognitive Science* Vol. 3, ed. Daniel Osherson, et. al. (Cambridge, MA: MIT Press, 1995), 385.

4. David Marr, *Vision* (San Francisco: W.H. Freeman, 1982).

5. Given limitations on time and memory, the multiplication problems one can actually solve is of course finite. For present purposes, we can safely ignore this complication.

6. Marr writes that we need a "level of understanding at which the character of the information-processing tasks carried out … are analyzed and understood in a way that is independent of the particular mechanisms and structures that implement them in our heads" (*Vision*, 19).

7. For useful discussions of the general topic of mental representation, see Braddon-Mitchell and Jackson, *Philosophy of Mind and Cognition*; *Representation in Mind Vol. 1: New Approaches to Mental Representation*, ed. Hugh Clapin, et. al. (Oxford: Elsevier, 2004); Martin Roth, "Representation, Philosophical Issues About," in *Wiley Interdisciplinary Reviews: Cognitive Science* 1 (1) (2010): 32–39.

8. Alan Newell and Herbert Simon, "Computer Science as Empirical Inquiry: Symbols and Search," *Communications of the Association for Computing Machinery* 19 (1976): 113–126.

9. Jerry Fodor, *The Language of Thought* (New York: Crowell, 1975).

10. Daniel Dennett, *Brainstorms* (Cambridge: MIT Press, 1978).

11. John Haugeland, "Representational Genera," in *Philosophy and Connectionist Theory*, ed. William Ramsey, et. al. (Hillsdale, NJ: Lawrence Erlbaum Associates, 1991), 61–89.

12. Andy Clark, *Being There* (Cambridge, MA: MIT Press, 1997).

13. Chris Swoyer, "Structural Representation and Surrogative Reasoning," *Synthese* 87 (1991): 449–508.

14. Robert Cummins, *Representations, Targets, and Attitudes* (Cambridge, MA: MIT Press, 1996); Ruth Millikan, *Varieties of Meaning* (Cambridge, MA: MIT Press, 2004).

15. Gabriele Contessa, "Scientific Representation, Interpretation and Surrogative Reasoning," *Philosophy of Science* 74 (1) (2007): 48–68.

16. Jerry Fodor, *Psychosemantics* (Cambridge, MA: MIT Press, 1987).

17. Donald Davidson, "Truth and Meaning," *Synthese* 17 (1967): 304–323.

18. H. P. Grice, "Meaning," *Philosophical Review* 66 (1957): 377–388.

19. For ease of exposition, I am using an uppercase word of English to talk about an expression in LOT. Obviously, it is not being claimed here that the would-be LOT expression that denotes cats is orthographically identical to the English word 'cat.'

20. Pierre Poirier and Robert Cummins, "Representation and Indication," in *Representation in Mind Vol. 1: New Approaches to Mental Representation*, ed. Hugh Clapin, et. al. (Oxford: Elsevier, 2004).

21. Cummins, *Representations, Targets, and Attitudes*.

22. For a nice illustration of the view that says skepticism about the propositional attitudes implies abandonment of RTM, see Louise Antony, "Everybody Has Got It: A Defense of Non-Reductive Materialism," in *Contemporary Debates in Philosophy of Mind*, ed. Brian McLaughlin and Jonathan Cohen (Boston: Blackwell, 2007).

23. Paul Churchland, "Conceptual Similarity across Sensory and Neural Diversity: The Lepore/Fodor Challenge Answered," *Journal of Philosophy* 95 (1) (1998): 5–32. For a more recent articulation of the view, see Paul Churchland, *Plato's Camera* (Cambridge, MA: MIT Press, 2012).

24. This view has been with us since Newton, and the framing of many major debates in twentieth-century philosophy of science—debates over scientific confirmation and explanation, for example—makes little sense without it (e.g., Ernest Nagel, *The Structure of Science* (New York: Harcourt Brace & Co, 1961); Rudolf Carnap, *Philosophical Foundations of Physics: An Introduction to the Philosophy of Science* (New York: Basic Books, 1966); Carl Hempel, *Philosophy of Natural Science* (New Jersey: Prentice-Hall, 1966).

25. Carl Hempel and Paul Oppenheim, "Studies in the Logic of Explanation," *Philosophy of Science* 15 (1948): 135–175.

26. Paul Oppenheim and Hilary Putnam, "The Unity of Science as a Working Hypothesis," in *Minnesota Studies in the Philosophy of Science* Vol. 2, eds. Herbert Feigl, et. al. (Minneapolis: Minnesota University Press, 1958), 3–36.

27. Jaegwon Kim, "Multiple Realization and the Metaphysics of Reduction," in *Supervenience and Mind* (New York: Cambridge University Press, 1993), 309–335; Jerry Fodor, "Special Sciences: Still Autonomous after All These Years," in *In Critical Condition* (Cambridge, MA: MIT Press, 1998), 9–24. Unlike Fodor, Kim challenges the conventional wisdom.

28. Jerry Fodor, "Special Sciences," *Synthese* 28 (1974): 77–115.

29. Elliot Sober, "The Multiple Realizability Argument against Reductionism," *Philosophy of Science* 66 (4) (1999): 550–551. Sober understands "unified explanation" in the sense of Philip Kitcher, "Explanatory Unification and the Causal Structure of the World," in *Scientific Explanation*, eds. Philip Kitcher and Wesley Salmon (Minneapolis: University of Minnesota Press, 1989). According to Kitcher, "Science advances our understanding of nature by showing us how to derive descriptions of many phenomena, using the same pattern of derivation again and again, and in demonstrating this, it teaches us how to reduce the number of facts we have to accept as ultimate" (423). Macro-explanations unify in that macro-explanations allow us to reduce the number of patterns used to explain phenomena.

30. Hilary Putnam, "Philosophy and Our Mental Life," in *Mind, Language, and Reality* Vol. 2 (New York: Cambridge University Press, 1975), 293.

31. Sober, "The Multiple Realizability Argument," 549.

32. Gualtiero Piccinini and Carl Craver, "Integrating Psychology and Neuroscience: Functional Analyses as Mechanism Sketches," *Synthese* 183 (3) (2011): 285.

33. It is of course possible that in some other system S*, or in S at different times, B is implemented, in which case it would be B, not A, that explains C.

34. Confirmation, as Fodor pointed out, appears to be isotropic and Quinian (*The Modularity of Mind* [Cambridge, MA: MIT Press, 1983]). By isotropic, Fodor means "that the facts relevant to the confirmation of a scientific hypothesis may be drawn from anywhere in the field of previously established empirical (or, of course, demonstrative) truths. Crudely: everything that the scientist knows is, in principle, relevant

to determining what else he ought to believe" (105). By Quinian, Fodor means "that the degree of confirmation assigned to any given hypothesis is sensitive to properties of the entire belief system; as it were, the shape of our whole science bears on the epistemic status of each scientific hypothesis" (107).

35. Piccinini and Craver, "Mechanism Sketches," 285.

36. William Lycan, *Consciousness* (Cambridge, MA: MIT Press, 1987).

37. Patricia Churchland and Terrence Sejnowski, "Neural Representation and Neural Computation," *Philosophical Perspectives* 4 (1990): 369.

38. Churchland and Sejnowski, "Neural Representation," 370.

Chapter 6

Fusion Finalized

If minds are things, then states of mind, i.e., mental states, are states of some thing. According to Cartesian materialism, the mind is the brain, so it would seem to follow that mental states are brain states. For example, if we think of possessing knowledge as a matter of being in a certain mental state, then it will be tempting to say that possessing knowledge is a matter of being in a certain brain state. By contrast, if having a mind is constituted by dispositions or capacities, as Ryleanism says, then the various "states" of mind are actually various dispositions, dispositions that collectively constitute "the mind." In a recent book,[1] Michael Pardo and Dennis Patterson provide a sustained critique of Cartesian materialism and seek to defend a version of Ryleanism. Of particular interest to Pardo and Patterson are the implications that conceptions of mind have for law, and they argue that many attempts to bring neuroscience to bear on questions of law rest on a problematic commitment to Cartesian materialism. While I am largely sympathetic to their project, Pardo and Patterson tend to write as if Cartesian materialism and Ryleanism are the only options available to physicalists. But Ryleanism has a number of shortcomings of its own, shortcomings that show up in Pardo and Patterson's discussion of the pertinent issues. Here I provide a critical examination of Pardo and Patterson's discussion, and I argue that functionalism provides a better alternative to Cartesian materialism than Ryleanism.

RYLEANISM AND ITS DISCONTENTS

According to Pardo and Patterson, "knowledge is a kind of cognitive achievement or success—it consists in a kind of power, ability, or potentiality possessed by a knowing agent."[2] The powers, abilities, or potentialities that constitute knowledge are behavioral dispositions, and Pardo and Patterson consider a challenge to this claim posed by patients who suffer from "locked-in syndrome." Such patients are said to be conscious and possess knowledge, despite the fact that they are unable to move or talk. In response, Pardo and Patterson write:

> Does this mean that knowledge is not, in fact, an ability, but rather something else (a brain state)? We think not. First, those with locked-in syndrome can, quite remarkably, learn to communicate their knowledge through a series of complex eye-movements. These communications do manifest knowledge consistent with an ability conception of knowledge. And before a locked-in sufferer learns to communicate in this way—or in cases of "total locked-in syndrome" in which no movements of the eye or any other body parts are possible—he is still able to reflect on his knowledge, to reason from what he knows to be so, and to feel emptions grounded in what he knows. These, too, are abilities or capacities and, indeed, these are reasons we ascribe knowledge to patients in this situation. If such patients were not conscious of their knowledge *in any way*, and could not manifest it in any way, on what basis would we ascribe knowledge to them? We would not.[3]

There are at least two problems with this reply, however. The first is that the abilities Pardo and Patterson attribute to locked-in patients—the abilities that presumably *constitute* their knowledge—are themselves characterized *in terms* of knowledge. For example, we are told that a locked-in patient is able to *reflect on* his knowledge, to *reason from* what he knows to be so, and to *feel emotions grounded in* what he knows. Each of these abilities presupposes knowledge, so these abilities cannot constitute the patient's knowledge. Similarly, Pardo and Patterson say that such patients can *learn to communicate their knowledge*, but this supposes a distinction between having knowledge and having the ability to communicate knowledge. But if knowledge is distinct from the abilities it enables—the ability to reflect on knowledge, to reason from knowledge, to feel emotions grounded in knowledge, and to communicate knowledge—then what are the abilities that constitute knowledge *itself*? Pardo and Patterson do not say, and it is not at all clear that there are any plausible candidates in the vicinity. A second, related problem is with Pardo and Patterson's use of "manifest." Pardo and Patterson frequently talk about manifesting, or showing, knowledge in behavior, but there are familiar uses of "manifest" and "show" in which to manifest or show knowledge is to give evidence of knowledge, in the sense of indicating knowledge. Just as red spots may provide evidence of chicken pox because red spots indicate chicken pox, behavior may provide evidence of knowledge because behavior indicates knowledge. But even if having red spots were the best evidence we could have for believing that a person has chicken pox, it wouldn't follow that having chicken pox is the *disposition* to have red spots. Chicken pox is the *cause* of red spots. Similarly, though it may be that behavior is typically the only feasible means for gathering evidence about what a person knows, it doesn't follow that knowledge is a disposition to behave. Though Pardo and Patterson claim that the evidential relationship between behavior and knowledge is "criterial" and not merely "inductive," sometimes they write as if it is only the inductive reading they are after. For example, in a footnote they admit that "someone may lose the ability to manifest his or her knowledge in some ways … or perhaps in all ways."[4] If the relationship between behavior and knowledge is merely inductive (in the sense of indicating), then it is perfectly coherent to say that a person has knowledge but is unable to manifest that knowledge in behavior, i.e., the behavior does not provide evidence that would justify a knowledge attribution. But if the relationship is "criterial," then what Pardo and Patterson say in this footnote clearly contradicts their earlier claim that possessing knowledge is a matter of having certain powers or abilities.

The two problems just noted resurface in Pardo and Patterson's discussion of whether information can be encoded in the brain. Pardo and Patterson agree with Maxwell Bennett and Peter Hacker, who claim that the brain does not possess knowledge in the sense in which a human possesses knowledge.[5] According to Pardo and Patterson, knowledge is an ability, not a state, and to support this claim, they ask us to imagine the following. Suppose that the schedule for the New York Philharmonic is "encoded" in your brain, but when you are asked questions about the dates of upcoming performances, you provide the wrong answers. If knowledge is a matter of being in a particular state, then we might be able to make sense of the claim that the relevant state is a brain state. In the example, the candidate brain state is one that encodes the New York Philharmonic schedule. But it is possible to be in such a brain state and yet fail to answer correctly questions about the schedule. If you fail to answer correctly questions about the schedule, we would rightly conclude that you didn't know the schedule.[6] Knowing is an ability, not a state; a fortiori, it's not a brain state. But if that's right, then it is obvious that brains don't know anything—brains lack the requisite abilities, i.e., brains do not possess the kinds of behavioral dispositions constitutive of knowledge.

Pardo and Patterson's argument here rests on a conflation of information, or representation generally, with knowledge. As we've seen, the appeal to information processing in cognitive science is done in service of explaining certain capacities and abilities, including, for example, your ability to correctly answer questions about the schedule. If we want to *explain* this ability, it will of course not do to say that you have this ability because your brain has this ability, for then we would want to know how your brain has this ability. For the sake of argument, let's *grant* that the relevant knowledge here is an ability. What explains it? The fact that this information is encoded in your brain. However, it is crucial that we make a distinction between possessing information and *exploiting* information. Consider my ability to successfully get around Chicago. This is explained by both the fact that I have an accurate street map *and* that I can use the map to guide my actions. If I were blind, the map would be of no use to me, but whether the map is accurate has nothing to do with whether I can use it. And if it is an inaccurate map, then I will fail to get around the city successfully, even if I know how to "read" maps.

The point of the above example is not that there is a little man in your head who reads an "internal" map in the very same sense in which you read a street map. *That* explanation would obviously be a nonstarter.[7] The point is just that there is a perfectly good sense in which a map can accurately represent even if no one takes advantage of that fact, e.g., if the map just sits there inert. The sense in which the map carries information, in the sense of information that includes representation, has nothing to do with abilities. Rather, it has to do with the structural similarity between the elements of the map and the street structure of the target (without which the question of accuracy makes no sense). Suppose we substitute the map for the schedule and grant that you cannot get around Chicago very well. Do you know your way around Chicago? No, for we are granting that knowledge is an ability and you lack it. But do you possess information about Chicago, in the sense of having an accurate representation of its streets? You do.

We need an alternative to Cartesian materialism and Ryleanism. The alternate picture, functionalism, emerges when we pay close attention to the practice of contemporary cognitive scientists. The explanatory targets are capacities, e.g., cognitive systems are intelligent, they are capable of learning, and they are goal directed, i.e., characterized by functions that are at least implicitly normative. But such a characterization of cognitive systems is the starting point of the science, and the explanations we find are built around the idea that such systems are information driven: to understand a cognitive system, you need to appreciate what information (and how much) it has about itself and its environment and/or containing systems. Some of the information is represented, i.e., encoded in objects or processes or states from which structural features of the things represented can be systematically recovered. Cognitive systems are also relatively complex, with the interactions among subsystems and components being themselves information driven.[8] And as I will show in a moment, this is precisely the picture we need in order to explain how reasons can be causes.

REASONS AS CAUSES, PART ONE

Because capacities are the explanatory targets of functional analyses, a physicalist-friendly functional analysis of free will and responsibility requires us to articulate, if only roughly at first, the way in which free will and responsibility can be characterized as capacities. In the previous chapter I characterized free will in terms of rationality, and I argued that critical to free will is the regulative role that the ideal of rationality plays in thought and behavior. Beings with free will are such that failures of rationality prompt changes in one's CAA, changes that bring one closer to D-rationality. I called this responsiveness A-rationality, and it should be fairly evident that to characterize someone as A-rational is to attribute to that person a certain capacity. Our task now is to consider what kind of system design would explain this capacity. Because the capacity has to do with responses to failure, such a system would have to be able to recognize both the norm relative to which the system's behavior counts as a failure and that a particular instance of behavior was in fact a failure. Each of these analyzing capacities can in turn be accounted for in terms of representation or information, i.e., the norm is the representational content of a belief about how one should act and the failure to satisfy the norm is the representational content of a belief or perception of how one acted on any particular occasion. However, though such cognitive and perceptual capacities may be necessary, it is clear that they are not sufficient. In order to bring the system closer to D-rationality, the system must also possess information about the cause of the failure and what actions the system can take to prevent such causes from being operative in the future. But even this is not enough, for as we've seen, a person can know that she failed to satisfy a norm, know how to prevent such failures in the future, and yet still not adapt her behavior in a way that brings her closer to D-rationality. In order for the failure to bring the person closer to D-rationality, the person has to *care* about her failures.

In Chapter 3 we considered the importance of the reactive attitudes to our experience as moral agents, and we noted that the force of certain demands, e.g., the goodwill demand, is reflected in the adoption of such attitudes toward ourselves and others. Cognitive scientists have not failed to appreciate the critical functional roles that such attitudes play. For example, Steven Pinker characterizes emotions as

> mechanisms that set the brain's highest-level goals. Once triggered by a propitious moment, an emotion triggers the cascade of subgoals and sub-subgoals that we call thinking and acting. Because the goals and means are woven into a multiply nested control structure of subgoals within subgoals within subgoals, no sharp line divides thinking from feeling, nor does thinking inevitably precede feeling or vice versa.[9]

Pinker illustrates this idea with the example of fear:

> Fear is triggered by a signal of impending harm like a predator, a clifftop, or spoken threat. It lights up the short-term goal of fleeing, subduing, or deflecting the danger, and gives the goal high priority, which we experience as a sense of

> urgency. It also lights up the longer-term goals of avoiding the hazard in the future and remembering how we got out of it this time, triggered by the state we experience as relief.[10]

These points apply to the "moral emotions" as well, i.e., the reactive attitudes. Drawing on the work of Robert Trivers, Pinker characterizes the moral emotions as strategies in a reciprocity game.[11] For instance, gratitude "calibrates the desire to reciprocate according to the costs and benefits of the original act. We are grateful to people when their favor helps us a lot and has cost them a lot."[12] And guilt is the reaction of "a cheater who is in danger of being found out … . People feel guilty about private transgressions because they may become public; confessing a sin before it is discovered is evidence of sincerity and gives the victim better grounds to maintain the relationship."[13] The thinking and acting triggered by the moral emotions is the thinking and acting that brings one closer to D-rationality; put another way, the moral emotions transform demands, including the goodwill demand, into effective reasons. This is their contribution to the containing system—their function—and when we understand the overall design of a system that includes such emotions, we understand how such a system can be A-rational.[14]

As Pinker's discussion of fear suggests, understanding how the moral emotions contribute to greater D-rationality requires us to appreciate the complex interplay between these emotions and other states. Fear prompts the short-term goal of fleeing, but it also prompts the long-term goal of avoiding such hazards in the future. Similarly, while guilt prompts the short-term goal of confessing and apologizing, it can also prompt the long-term goal of avoiding future transgressions. Whether this latter goal is met, however, may depend on the presence of other affective states, states whose presence is not automatically guaranteed by the original feeling of guilt. The significance of this point is illustrated by the well-known case of E.V.R., a man who suffered damage to his prefrontal cortex. In a widely discussed study of patients with such damage, Antoine Bechara, Antonio Damasio, and their colleagues write that

> Patients with damage to the ventromedial sector of prefrontal cortices develop a severe impairment in real-life decision-making, in spite of otherwise preserved intellect. The impairments are especially marked in the personal and social realms … . E.V.R. is a prototypical example of this condition. He often decides against his best interest, and is unable to learn from his mistakes. His decisions repeatedly lead to negative consequences. In striking contrast to this real-life decision-making impairment, E.V.R.'s general intellect and problem-solving abilities in a laboratory setting remain intact … he is not perseverative, nor is he impulsive; his knowledge base is intact and so is his short-term and working memory as tested to date; his solution of verbally posed social problems and ethical dilemmas is comparable to that of controls.[15]

Bechara and Damasio describe an experimental task that simulates real-life decision-making in a way that reflects the uncertainty of assumptions and outcomes, including reward and punishment, often present when a person is faced with a decision. In the experiment, subjects are presented with four decks of cards that are equal in size and appearance, and subjects are asked to draw cards (one at a time) from any of the four decks. Subjects receive money after each draw, but after some draws subjects are required to pay a penalty. Each subject receives an initial loan of play money, and the goal of the task is to maximize profit from that loan. The decks differ in an important respect, however. While the payouts for decks C and D are on average significantly lower than the payouts for decks A and B, the penalties are also less severe. In the long run, drawing from only decks C and D results in a net gain; by contrast, drawing from only decks A and B results in a net loss. The results of the experiment showed that normal controls made more selections from the good decks (C and D), and avoided the bad decks (A and B). Unlike controls, patients with prefrontal damage (including E.V.R) selected fewer from the good decks and chose more from the bad decks. The difference was significant, and repeated testing did not alter E.V.R.'s performance on the task. By contrast, the performance of normal controls improved over time.

Bechara and Damasio conclude that subjects with prefrontal damage "are unresponsive to future consequences, whatever they are, and are thus more controlled by immediate prospects."[16] However, these patients seem to possess the knowledge needed to imagine various courses of action and the future consequences of such actions, and thus "[t]heir defect seems to be at the level of acting on such knowledge."[17] Bechara and Damasio hypothesize that while such patients can construct stable representations of future outcomes, such representations are not "marked" with a negative or positive value. This "somatic marker hypothesis" is further explored in Damasio's book *Descartes' Error*,[18] and the basic idea is that representations of the consequences of considered actions typically have an affective component to them, and it is the affective component of such representations that biases decision-making one way or another. Damasio argues that the ventromedial frontal region of the brain plays a critical role in integrating the cognitive and affective components of such a representation, components made available via connections to the amygdala and the hypothalamus. Because such connections are disrupted in E.V.R., he does not act on what he "knows" to be the best course of action. Put another way, while E.V.R. can generate the representations of consequences of future actions that constitute reasons for acting one way or another, the affective component necessary for making those reasons effective is missing.

A number of take-home points emerge from these studies. The first is what while the results and hypotheses offered by Bechara and Damasio may be challenged in various ways, the methods and strategies used to approach the subject are the bread and butter of cognitive science. The second is that a close examination of Bechara and Damasio's work reveals that the methods and strategies used are a specific instance of the explanatory framework described by Marr. There is a specification of a capacity: rational decision-making; there is a functional analysis that appeals to information processing: constructing representations of the consequences of considered actions; there is a specification of the physical mechanisms that implement the functional analysis: the ventromedial frontal region and its connections to the amygdala and hypothalamus. Reflection on this second point prompts a third. While Bechara and Damasio might embrace the spirit of *you are your brain*, nothing in their work suggests that they would say *rational decision-making is a brain state*. Rationality has to do with behavioral capacities, just as Pardo and Patterson insist. Bechara and Damasio's interest is in *explaining* those capacities, but even here, the explanatory relation between brain and behavioral capacity is not direct. Instead, the relation between brain and behavioral capacity is mediated by a functional analysis of the capacity, an analysis that appeals to information processing (among other things). For Bechara and Damasio, the phrase *you are your brain* merely expresses the conviction that the functional analysis of the capacity—the analysis that *explains* the capacity, in the sense of functional explanation—is implemented in the brain.

The methods and strategies exemplified in the study of E.V.R. provide us with the resources needed for showing how reasons can be causes, and thus showing how the manifest and scientific images can be fused. To see what fusion looks like, it will help to first consider the alternative to fusion provided by Pardo and Patterson, an alternative that is predicated on rejecting the explanatory strategy that is at the heart of contemporary cognitive science. Recall Pardo and Patterson's example of a person stopping a car at a red light, discussed in Chapter 1. Though a traffic light may "cause" the person to apply the brakes, Pardo and Patterson deny that the light causes the person to apply the brakes in the same way that a bowling ball causes pins to fall over. Unlike pins, we act for reasons and have choices.[19] But it is important to distinguish three different issues here. The first is whether the relationship between the illumination of the red lamp and the pressure applied to the brakes is like the relationship between the bowling ball striking a set of pins and the pins falling over. In one obvious sense the answer is no, for unlike the relationship between the bowling ball striking the pins and the pins falling over, the relationship between the illumination of the red lamp and the pressure applied to the brakes is *mediated* by a complex sequence of events (events that occur in the person's brain and body), i.e., the causal connection is indirect in a way the causal connection in the ball and pins case is not. As such, any explanation that does not make mention of this intervening sequence of events will leave something important out and thus look incomplete.

This leads to the second issue, which is whether an explanation of the event "stopping at a red traffic light" can be given in purely physical terms. Pardo and Patterson's suggestion is that even if we gave a complete account of the physical goings on, i.e., even if we included all the intervening physical events that occur in the person's brain and body, we would not have an explanation of the target event. What the purely physical story leaves out is the fact that we have a reason for stopping, and since it is this fact that explains our stopping, the purely physical story does not provide the kind of understanding we seek. We've encountered a similar idea already, in our discussion of how we might explain a calculator's capacity to multiply. The point there was that a person could have complete knowledge of the physics of the calculator's circuit board and yet not know that the calculator executes a partial products algorithm. Insofar as the kind of understanding we seek is provided by the partial products algorithm, the physics of the circuit board does not really explain the calculator's capacity to multiply. As we saw, the problem is that, unlike the description of the algorithm, the physics does not group events into the functionally salient units that ground the explanation, and thus while in any particular computing environment executing an algorithm *just is* a matter of going through a sequence of physical states, we need the causal relevance filtering provided by the algorithmic description to "see" the sequence of these states as the execution of the partial products algorithm.

The third issue is whether reasons can be causes. Pardo and Patterson deny that reasons are causes, for "Unlike the pins, we choose whether to stop at the red light. If we fail no stop, we run the risk of sanction. The pins in a bowling alley have no such choice: they are 'compelled' to fall over by force of the impact of the bowling ball."[20] The argument seems to be this: though the red light provides a reason for stopping and this reason explains why I do in fact stop, the reason does not compel me to stop. By contrast, physical causes compel the events they are said to cause. Thus reasons are not causes in the way physical events are causes. While this may seem like an intuitively plausible line of reasoning, it faces a serious difficulty. On the one hand, Pardo and Patterson do not seem to have any objection to saying that the behavior of the pins is compelled by the force of the bowling ball, and if by "compel" we mean "determine" and if we assume that the example of the bowling ball applies to physical events generally, then Pardo and Patterson do not seem to have any objection to saying that the physical world is a deterministic world. But if stopping at a red light is supposed to be importantly different from the behavior of the pins, and if the key point about the behavior of the pins is that this behavior is compelled—that is, determined—then this suggests that stopping at the red light is *not* determined. If physical events are determined, however, then it also follows that stopping at the red light is not a physical event. To avoid this absurd conclusion, we must either accept that reasons can be causes or reject determinism. If we reject determinism, however, we lose the only basis Pardo and Patterson offer for distinguishing physical events from human actions, i.e., we can no longer say that the behavior of the pins is compelled whereas stopping at the red light is not. In the same vein, if we accept that reasons can be causes, then human actions are not importantly different from the behavior of bowling pins; each is "compelled."

Pardo and Patterson are correct that we can be responsible for our actions, and they are correct that bowling pins are not responsible for their behavior. There is also nothing wrong with using the language of "choice" and "reasons" to capture this difference. As we have seen, however, the significance of *having reasons* to the issue of responsibility has nothing to do with whether behavior is compelled, where by "compelled" we mean something like "determined." The problem with assimilating human behavior to the behavior of bowling pins is not that the latter is compelled but the former is not; rather, the problem is that the causes of human behavior have features that the causes of bowling pin behavior do not, features that we will miss if we try to explain human behavior using exclusively the vocabulary appropriate to bowling pins. Pardo and Patterson come close to appreciating this point when they note that purely physical explanations leave out the "intentional element" in the explanation of human behavior,[21] but their subsequent discussion unnecessarily casts the intentional element in metaphysical terms. Physical explanations differ from intentional explanations, but not because "physical" and "intentional" denote different metaphysical categories. Unless you are a dualist, the intentional is a proper part of the physical. But being physical doesn't imply that the vocabulary of physics, or even neuroscience, is the only—or best—vocabulary for couching explanations. With that said, the metaphysics does put some constraints on the explanatory vocabularies available to us, and unless we are able to put those vocabularies in some relation to each other, we may reasonably worry whether these vocabularies are actually consistent with physicalism. One constraint is that if your conception of physical reality is deterministic and you reject mind-body dualism (as Pardo and Patterson do), then any viable explanatory vocabulary had better not introduce nondeterministic elements. This is precisely where Pardo and Patterson's account runs into trouble. But even if intentional explanations need not violate this constraint, there remains the burden of relating our intentional explanations to our neuroscientific ones. In terms introduced in Chapter 5, a satisfying account of mind and brain requires providing both horizontal explanations and vertical explanations, i.e., it requires us to introduce reasons into our framework without making them explanatorily idle or metaphysically suspect. Bechara and Damasio's work on E.V.R. provides the outlines of what such an account would look like.

So how do we fit the "intentional element" into the explanation of human behavior? Recalling our earlier discussion of reasons and the account of representation presented in the previous chapter, we can treat reasons as psychological states that have the right kind of content and play the right kind of functional role. Pardo and Patterson dismiss this option in a footnote, claiming that "Strictly speaking, reasons are not efficient (i.e., mechanical) causes."[22] But this claim is doubly problematic. First, while we can agree with Pardo and Patterson that *normative* reasons are not causes, there is a tradition tracing back to Aristotle that identifies explanatory reasons with "efficient" or "mechanical" causes, and as we saw, explanatory reasons can be normative reasons in the derivative sense that an explanatory reason can have a normative reason as its content. Pardo and Patterson apparently think this tradition is untenable, but they don't mention it, much less refute it. Second, in the typical sense of "mechanical," an explanatory reason can be an efficient cause even if it is not a mechanical cause. For example, William Lycan notes that "common sense … does not treat beliefs as billiard balls or levers … a belief is not portrayed by folk wisdom as a *mechanical part* of a person. It is not that Erica's belief *pushed on* part C, which activated engine E, and so on."[23] But denying that beliefs are mechanical parts of a person does not entail that beliefs are not causes, and Lycan uses the example of computers to illustrate the point:

> For example, a conventionally programmed home computer stores sentences (say, the English sentences being manipulated by a word processor) in a physically scattered way, utilizing widely disparate memory addresses inscrutably linked; if one asks even an expert technician *where* the computer has stored a particular sentence (such as "With a gasp of relief, Reg unstrapped the slab of raw liver from his left shin"), the technician will have not the faintest idea, and will doubt that the question has any crisply factual answer. Yet the sentence is stored in a *functionally* discrete way; a few keystrokes' worth of input can perform a teleologically defined word-processing operation on that very sentence *qua* sentence, such as moving it, underlining it, checking its spelling, or making it a separate paragraph.[24]

This is an example of a "physically scattered but functionally discrete representation,"[25] and "a scattered representation's lack of a palpable, robust cog-like causal role is by itself no embarrassment to the … identification of a belief with it."[26] Beliefs are thus an instance of what Andy Clark calls a scattered cause, where scattered causation occurs "when a number of physically distinct influences are usefully grouped together … and are treated as a unified force for some explanatory purpose."[27] Perhaps Pardo and Patterson would find this response problematic, but here, too, they do not mention the possibility, much less refute it.

REASONS AS CAUSES, PART TWO

Making good on the claim that reasons can be causes requires us to clear one more conceptual hurdle, however. In Chapter 2 I noted the argument to the effect that role functionalism renders psychological explanations of human action vacuous. For current purposes, we can treat role functionalism as a commitment to the kind of nonreductive physicalism I defended in the previous chapter. If psychological explanations include reasons explanations, reasons explanations are causal explanations, and causal explanations are not vacuous, then the defender of nonreductive physicalism needs to show why the argument against role functionalism fails. I've said that beliefs and desires are among the explanatory reasons for action, and the following principle of folk psychology (FP) provides a general schema that captures the way that beliefs, desires, and action are related:

> [L] If any person, agent, individual, wants some outcome, *d*, and believes that an action, *a*, is a means to attain *d* under the circumstances, then *x* does *a*.[28]

Superficially, [L] looks like the kind of law statement familiar in science: just as Boyle's law reveals how temperature, pressure, and volume are connected, and Ohm's law reveals how current, voltage, and resistance are connected, [L] reveals the manner in which beliefs, desires, and actions are connected. As such, [L] appears to be at least a candidate law in a mature science of psychology. If beliefs and desires (typically) provide our reasons for action, then whether an action can be located in both the space of reasons and the space of causes turns on whether [L] can be treated like other law statements we find in science.

The thought that a generalization like [L] would be needed in order for folk psychological explanations of human action to be scientifically legitimate derives largely from a commitment to the deductive-nomological account of explanation discussed in Chapter 5. Recall that according to the deductive-nomological account, individual events are explained when they are deduced from laws—understood as universal generalizations—and a statement of initial (or boundary) conditions. Applied to the present case, the events to be explained are actions, and an action is explained just in case it can be deduced from [L] plus initial (or boundary) conditions. The availability of a generalization like [L] is not enough to establish that folk psychological explanations are on all fours with other explanations in science, of course, for Boyle's law and Ohm's law appear to state contingent connections. However, the contingency of folk psychological generalizations seems to be secured by the thought that folk psychological explanations cite causes, and causal connections are contingent. So, for example, Jerry Fodor writes, "causal powers are a function of your *contingent* connections, not of your conceptual connections. As, indeed, Uncle Hume taught us."[29] Let's call the dual commitment to a deductive-nomological account of explanation and a Humean account of causation the *Hempel-Hume* account of explanation (hereafter "H-H").[30]

However, if legitimate scientific explanations are H-H explanations, then there are serious grounds for doubting that principles like [L] can provide legitimate scientific explanations. The problem comes into focus when we consider the familiar, widely accepted functionalist interpretation of FP. For example, in their well-known defense of FP, Terence Horgan and James Woodward write:

> Folk psychology is a network of principles which constitutes a sort of commonsense theory about how to explain human behavior. These principles provide a central role to certain propositional attitudes, particularly beliefs and desires. The theory asserts, for example, that if someone desires that *p*, and this desire is not overridden by other desires, and he believes that an action of kind *K* will bring it about that *p*, and he believes that such an action is within his power, and he does not believe that some other kind of action is within his power and is a preferable way to bring it about that *p*, then ceteris paribus, the desire and the beliefs will cause him to perform an action of kind *K*. The theory is largely functional, in that the states it postulates are characterized primarily in terms of their causal relations to each other, to perception and other environmental stimuli, and to behavior.[31]

As we've seen, functionalism says that what makes some state a state of, e.g., belief, is the *role* that state plays in the larger cognitive economy of an organism. Fodor himself says that the generalization "if you want that P and you believe that not-P unless Q, then all else being equal, you try to bring about that Q" is the kind of generalization that functionalists should be prepared to accept, and Fodor happily concedes that a belief state may be *defined* as one that causally interacts with desires and actions in the way specified by the generalization.[32] But if generalizations like [L] inter-define belief, desire, and action, [L] can hardly be a *contingent* law statement that is subject to empirical testing. Rather, if our commonsense mental concepts are broadly functional concepts, then certain connections between mental states and actions are conceptually guaranteed. Robert Cummins expresses this point nicely, noting that "When we describe causal interactions between functionally characterized components, the relevant causal generalizations pretty much come for free, since a functionally characterized component is a component identified by its relevant causal powers."[33] Insofar, then, as there are folk psychological *laws*, these laws trivially follow from functionalism about the mental. The moral of the story is clear: *functionalist folk psychology is incompatible with H-H.*

This moral hasn't been lost on certain critics of functionalism. For example, Robert Rupert relies on the incompatibility to argue that functional explanations in psychology are vacuous:

> The problem comes into focus more clearly when we consider the law of a supposedly autonomous empirical psychology meant to underwrite the single-case explanation: A subject that is in some state or other that causes, among other things, aversion behavior exhibits aversion behavior. Such a "law" appears vacuous, stating no more than that states that cause *e* cause *e*. Metaphysical necessity—not empirical law—grounds this claim.[34]

Here we find a clear commitment to H-H: science is (or should be) in the business of providing causal explanations of individual events by subsuming those events under contingent, empirical laws. The "laws" that functionalism generates, however, are trivial. Thus, if folk psychological concepts are functional concepts and [L] expresses (in part) the functional roles of belief and desire, then [L] cannot factor into the kind of explanations required by H-H.

Rupert's rejection of functionalism starts with the incompatibility of functionalism and H-H, and because he affirms the latter, the former has to go. But function talk is rife in science, engineering and everyday life, and terms like "gene," "catalyst," "resistor," "brain," "heart," and "force" are all, in some of their uses, functional. If Rupert is right, however, it appears that the functional explanations in which these terms figure are all vacuous. Something has gone seriously wrong here, but the culprit is H-H, not functional explanation.

Increasingly, philosophers have emphasized the centrality of *models* in science,[35] and in his seminal *Explaining Science*, Ronald Giere argues that law statements (or "principles," as Giere prefers to call them) are best thought of as specifying models, rather than making direct claims about the world.[36] As Giere points out, if we interpret law statements as asserting generalizations about the world, all of our laws turn out to be literally false. Philosophers of science have recognized this, of course, and some have tried to overcome the problem by articulating a notion of approximate truth.[37] However, given the well-known problems with providing a clear and coherent notion of approximate truth,[38] and given that models can do the required representational work, Giere advises us to give up trying to salvage approximate truth. Rather, we should regard law statements as true "by definition" of the models they partially specify, with the empirical content of science consisting in the theoretical hypothesis that a particular model is *similar*, in degrees and respects, to some target domain. According to Giere, most of the models we find in science are *abstract entities* that possess all and only the properties that we characterize them as having, and instead of being true or false, these models are more or less accurate representations of the domains we assert them to represent. In the spirit of Giere, we can say that principles like [L] function to specify *models of intentional systems*.[39]

Giere's approach to understanding how conceptually guaranteed laws contribute to empirically grounded explanations emerges from his discussion of the role $F = ma$ plays in Newtonian mechanics. Noting the long-standing dispute over whether this law should be construed as an empirical generalization or a definition, Giere (1999) claims that it should be understood as a principle for building models, models that can be brought into agreement, closely enough for our purposes, with other perceptual or theoretical phenomena.[40] Furthermore, very general principles like $F = ma$ do not specify models by themselves. Rather, a model is constructed from $F = ma$ when we specify the relevant forces. When the forces are specified, we end up with a model of, e.g., a linear oscillator. Paul Teller's discussion of Hooke's law provides a nice example of the idea. Teller writes:

> Hooke's law provides a simple example of what I have in mind. To what does it apply? What counts as a spring? Whenever the restoring force for a deformed material can be expressed as a Taylor series there is a linear first term. We count a material as a spring just in case, for ranges in which we are interested, the restoring force is well approximated by the linear first term of the Taylor expansion, that is, exactly if the material satisfies Hooke's law.[41]

Applied to [L], the empirical question is not whether [L] is a true, contingent generalization about the world, but whether anything satisfies a model specified using [L]; to the extent that humans exhibit the effects predicted by such a model, to that extent we can regard humans as an intentional system of that kind. For present purposes, however, the key point is that if the failure to state a contingent generalization renders [L] explanatorily vacuous, then Hooke's law is equally vacuous: after all, if we *count* a material as a spring just in case the restoring force is well approximated by the linear first term of the Taylor expansion, it turns out that "springs satisfy Hooke's law" isn't a contingent generalization, either. To understand why this does not entail that models specified by Hooke's law—or [L]—are explanatorily empty, we need to clarify *what* and *how* a good many models explain.

According to Giere, science isn't primarily in the business of explaining events. This isn't to say that scientists aren't concerned with explaining events, of course, but only that this concern is largely secondary. To see what Giere has in mind, consider the physiologist who attempts to explain bodily capacities and effects by developing models of various systems, e.g., the nervous system or the circulatory system. A physiologist might develop a model that is highly accurate of some system in our body without explaining, of any particular event that occurs in our body, why it occurred. Her goal is not to explain this, but to construct a model of how our bodies work—what the various parts are, how the interaction of those parts reliably produce certain effects, and so forth. As we've seen, many of the explanations we find in science, engineering, and everyday life are like this: instead of explaining particular events, these explanations consist in providing a functional analysis of a capacity or dispositional property. For example, imagine an electrical engineer's explanation of how a particular lighting system works. The engineer would specify a *model* of a circuit, a model that includes various components (e.g., resistors, power supplies) and how those components interact. In the model, whether a component is a resistor depends only on whether the component causes potential drops, and so "resistors cause potential drops" will be true of the model by definition. The *explanatory* payoff of such a model is that, because components are characterized solely in terms of contributions relevant to analyzing the target capacity, it becomes transparent why any system satisfying the model will have the target capacity.

The worry that "analytic" generalizations like "resistors cause potential drops" cannot figure into genuine explanations is an artifact of H-H's commitment to a deductive-nomological account of explanation. If we accept that explanation is subsumption under law, then we will be tempted to think that the job of a generalization like "resistors cause potential drops" is to explain particular occurrences of potential drops (events). According to H-H, the generalization cannot perform this job if the generalization is not contingent—if a resistor just *is* something that causes potential drops, any explanation of potential drops that appeals to "resistors cause potential drops" would look vacuous. However, reflection on the circuit model should make it clear that the worries about contingency are misplaced, or at least premature. Functional analyses are principally geared toward explaining *capacities*, not events, so from the outset H-H misidentifies the primary explanatory targets of functional analysis. The role that the generalization "resistors cause potential drops" plays isn't that it accounts for ("covers" or "subsumes") particular potential drops (individual events). By functionally characterizing components as *resistors*, we isolate effects of components in a way that makes them (potentially) relevant to an analysis of some capacity. The generalization makes explicit what those contributions are, of course, but it's the models in which functionally characterized components figure that are doing the explanatory work.

If I know that a circuit model accurately represents some aspect of a physical system and know that a part, X, of the physical system corresponds to the resistor, then I am in a position to understand why, if a current passes through X, the characteristic effects of such a system occur, e.g., the lights go out. The model provides the understanding of how certain aspects of this physical system work, and identifying part of the target as a resistor, e.g., using "this resistor" to pick out a nickel-chrome wire, reflects an assumption that the circuit model is an accurate representation of certain aspects of this physical system (target). But the model itself does not guarantee that the nickel-chrome wire is a resistor, or that a potential drop occurs when a current passes through this wire, or that the lights go out when there is such a potential drop. The analyticity or metaphysical necessity is in the model, so to speak, and not the target. As far as explanation goes, the appeal to the model is thus both *indispensable* and *non-vacuous*. These points apply, mutatis mutandis, to springs: in terms of the *model*, a spring is anything having a restoring force that is well approximated by the linear first term of the Taylor expansion, and so "springs have a restoring force that is well approximated by the linear first term of the Taylor expansion" is true by definition. However, specifying components in a schematic diagram (model) as *springs*, i.e., as things that satisfy Hooke's law, may be precisely what is needed to understand how something, e.g., my child's pogo stick, works. To the extent that the model is accurate of some target system, this licenses calling an aspect of the target—a piece of coiled metal in my child's pogo, for example—"spring." However, because the license rests on a *contingent* relation between model and target, the explanatory gains provided by the model do not entail that the analyticity or necessity *in the model* transmits *to the target*, i.e., it does not entail that "this piece of coiled metal satisfies Hooke's law" is analytic or necessary.

Linking elements of a model to the perceptual and theoretical phenomena we are trying to represent is also crucial to evaluating the *accuracy* of models, of course. In the examples of the lighting system and pogo stick, we can imagine that the model builder has a target system in mind when specifying a model. But a model of a lighting system, or a pogo stick, is not itself a lighting system, or a pogo stick; bringing a model to bear on some target system requires *identifying* aspects of the model with aspects of a target system.[42] How, then, can we test whether a specified model accurately represents a particular target system? Two factors are relevant to such a test: (i) the extent to which the model predicts the effects or regularities the system is known to exhibit, and (ii) the extent to which the functional characterizations employed by the model can be shown to be implemented in the target system. To illustrate the general idea here, imagine that we want to explain water runoff patterns in a certain region. Generate a topographic map of the region, superimpose assumed rainfall quantities and locations on the map, and run a simulation. If the result of the simulation matches our runoff data, then this is evidence that our topographic map is accurate. We can also use the map to make novel predictions about runoffs by superimposing possible rainfall quantities and locations. If the predictions about runoff prove correct, this provides additional support for thinking the map accurate. If our simulations don't work, this might be the result of incomplete or inaccurate information about initial rainfall quantities and locations. If that is not the problem, the map can suggest likely places where the map is inaccurate or incomplete. This hypothesis can be tested, and along with it the map itself. In any case, it is important to notice that the relations between the map's components are "analytic," or certainly not contingent; it is just geometry, after all. The contingency is in the fit between map and world (model and target), i.e., it is contingent whether the geometry actually is, or approximates, the spatial properties of the target. Compare: Kepler's equal areas law is analytic in Euclidean space, but it is contingent (and actually false, but a very good approximation) that it holds of the planets in our solar system.

With these points in mind, consider Horgan and Woodward's claim that "It is satisfaction of the 'causal architecture' of FP, by some set of (possibly complex) events in the central nervous system, which is crucial to the truth of FP."[43] If we think of the causal architecture of FP as the content of a model of an intentional system (where [L] is true by definition of the model) and treat aspects of the central nervous system as our representational target, then the "truth" of FP turns on the fit, in specified degrees and respects, between the model and the central nervous system. The task of *confirmation* is thus the task of identifying elements and functional relations in the model with certain states of the central nervous system and causal relations among those states, and we can test these models in much the same way we can test topographic maps. The *explanatory* power of FP resides largely in the extent to which principles like [L] can be used to construct a family of models, models that can account for the wide range of behavioral capacities of humans. As with "resistors cause potential drops" and "springs have a restoring force that is well approximated by the linear first term of the Taylor expansion," that [L] is true by definition of these models reveals that we are identifying states solely in terms of specific contributions those states make to a system. Again, this is precisely what's needed in order for a functional analysis of some capacity to succeed, and if the price of this explanatory transparency is a trivialization of the resulting generalizations, this is a price well worth paying. To repeat: because the generalizations are true by definition of models, not targets, and because these analyses do not subsume events under laws, trivializing the generalizations does not render the models explanatorily vacuous.

Functional explanation of the sort familiar from psychology and biology typically consists in providing a functional analysis of a capacity or property, e.g., the Planning Effect, with the analysis provided constituting an abstract model of the capacity or property. The explanatory power of an analytic strategy resides largely in the extent to which it can be used to construct a family of models, models that can be brought into agreement with other theoretical or perceptual phenomena. The history of cognitive science demonstrates Peter Godfrey-Smith's claim that "Cognitive science encounters the folk psychological model as a starting point and as a source of structural ideas."[44] In effect, early cognitive scientists were using principles like [L] to construct models of intentional systems, and though competing ideas about how to develop the details emerged, these models were unified insofar as each was constructed (tacitly or explicitly) in accordance with principles like [L]. Included among the challenges facing contemporary cognitive science are both figuring out the extent to which new theoretical models (e.g., those inspired by work in robotics, dynamical systems theory, connectionism, and neuroscience) can be adequately viewed as based on principles like [L] and, if they cannot be so viewed, determining whether these new models should be thought of as rivaling or complementing existing [L]-based models.

With these points in mind, let's return to the illustration of the Planning Effect (Chapter 5, Figure 5.1). Some elements and operations are described using the language of belief and desire, and if we had some way of characterizing belief and desire independent of the contributions they make (represented in the diagram), then we might think the diagram reflects a contingent, empirical discovery. However, if being a belief (or desire, or whatever) *just is* being something that plays a certain role in a system, e.g., the role depicted by the figure, then it seems the diagram isn't really *informative*. The worry is misplaced, however. For starters, we can eliminate the use of "belief" and "desire" in our description of elements and operations, e.g., we can replace "belief" with "box 1" and "desire" with "box 2." But notice that the contributions these boxes make to the system, as represented by the diagram, are just the contributions needed to understand why the system has the capacity we are trying to explain. In other words, the diagram *leaves out* many of the causal consequences of the boxes. So how can we pick out the relevant contributions and ignore the irrelevant ones? By characterizing the boxes as beliefs and desires! To treat something as an intentional system is to treat it as something that has the kind of design depicted in the figure. Here is another way to make the point. Suppose the component labeled "belief" is relabeled "box 1," and suppose that in relabeling it this way we remove all causal relevance filters. If I want the diagram to represent the causal consequences of the component, then I am going to have to add a lot more arrows to the diagram. But if I do that, then I lose the understanding provided by the original diagram. Having used "belief" to isolate the relevant contributions, I can then of course replace "belief" with "box 1," for now the information that the system has beliefs is already captured by what the diagram includes and what it leaves out. Again, the explanation is provided by the design, and to be an intentional system is to be a system that has this kind of design. Thus if a particular system, S, in fact implements this design, then S is an intentional system.

Our original worry was over whether reasons can be causes, and we framed the issue as the issue of whether an episode can be located in both the space of reasons and the space of causes. Locating an action in the space of reasons is tantamount to using a principle like [L] to explain the action, and locating an action in the space of causes seems to require that [L] be a contingent generalization. Apparently [L] isn't contingent if functionalism is true, so it looks like functionalism blocks our attempt to locate an episode in both the space of reasons and the space of causes. But there is a plausible reply available to the functionalist, one that is motivated both by considerations of the role of laws in scientific explanations generally and by reflection on the explanatory strategy of functional analysis. When we explain a person's actions in terms of her reasons, we treat this person as an intentional system. If a system is an intentional system, then the system implements a design that is specified in terms of the organization of the functionally characterized components *belief* and *desire* (and so forth). Because the system implements this design, there are components of the system that contribute what is specified by the functional characterization of belief, and of desire, and this is what licenses our calling a component a belief, or calling a component a desire. Given the way the design is specified, [L] (or something close to it) will look to be "true by definition" of any system that satisfies an intentional model.[45] But even if we grant this, it won't be true by definition that any particular system implements the design, and if a system does implement the design, it won't be true by definition that a particular component contributes what is specified by the functional characterization of belief, or of desire (and so on).[46] In that sense, [L] is clearly contingent.

What then of role and realizer functionalism? Because a component's status as a belief is itself explained in terms of the contribution the component makes to the system, we cannot explain why a component makes the contributions it does by saying it is a belief, i.e., it is not the second-order property of being a belief that causes anything to happen. In this way, realizer functionalism is right that the first-order properties of the component do all the causal work. In the context of functional analysis, the value of functional characterization is not that it picks out properties in virtue of which something has causal powers, but rather that it picks from among a component's causal powers those that are relevant to the explanation of some capacity. That is the important kernel of truth in role functionalism, one that gets lost if we wed functional explanation to the causal explanation of events. Once we get clear about this, we are able to understand in what sense reasons can be causes. Consider an occasion in which a system *exercises* one of its capacities, e.g., the system performs some bit of behavior. Because the capacity is explained in terms of a design, the exercise of the capacity will be explained in terms of what the components, specified in the design, are doing. If the causal contributions of these components include those picked out by the functional characterization of *belief*, and of *desire* (and so forth), then the causal explanation of the behavior will involve the very features on which having a reason for the behavior depends. If this does not amount to locating an episode in both the space of reasons and the space of causes, then it is not clear what would.

It may seem that my defense of fusion commits me to the integrity of FP and the propositional attitudes, and if it does so commit me, then it might be objected that defending fusion requires responding to the arguments of eliminative materialists. For example, Paul Churchland argues that neuroscience will likely show FP to be a false theory of the inner causes of human behavior,[47] and because I've characterized fusion as an alternative to fission and elimination, it looks like I have to show what is wrong with Churchland's argument. But appearances to the contrary, Churchland's brand of eliminative materialism does not threaten fusion. To appreciate why this is so, consider Pardo and Patterson's claim that "The reductionist wants to eliminate the intentional element (i.e., reasons for action) in the explanation of behavior. In other words, the reductionist wants all explanation to be in terms of causes and not rules or reasons."[48] If Pardo and Patterson count Churchland as a reductionist, as they do,[49] then their claim here is simply false. In rejecting the propositional attitudes, Churchland is not rejecting intentionality; rather, as we saw in the last chapter, Churchland is replacing the account of intentionality provided by the language of thought (LOT) with the account of intentionality provided by state-space semantics (SSS). Normative reasons are facts or states of affairs, and Churchland has no objection to saying that brains represent facts or states of affairs. His objection is rather to the claim that human representation is largely or exclusively lingua-form, but this will provide grounds for eliminating intentionality only if all representation is lingua-form. Churchland clearly does not argue this way, however. The account of fusion I defend appeals to intentionally characterized states and processes that play certain functional roles in a person's CAA, and while I have typically framed the account in terms drawn from FP, the requirements for fusion can be met within Churchland's preferred framework.

NOTES

1. Pardo and Patterson, *Minds, Brains, and Law*.
2. Pardo and Patterson, *Minds, Brains, and Law*, 18.
3. Pardo and Patterson, *Minds, Brains, and Law*, 19.

4. Pardo and Patterson, *Minds, Brains, and Law*, 22.

5. Maxwell Bennett and Peter Hacker, *Philosophical Foundations of Neuroscience* (Malden, MA: Blackwell, 2003).

6. Pardo and Patterson, *Minds, Brains, and Law*, 21.

7. I return to this point in Chapter 8.

8. Robert Cummins, et. al., "Epistemological Strata and the Rules of Right Reason," *Synthese* 141 (3) (2004): 287–331.

9. Steven Pinker, *How the Mind Works* (New York: W.W. Norton & Company, 1997), 373.

10. Pinker, *How the Mind Works*, 374.

11. Robert Trivers, "The Evolution of Reciprocal Altruism," *Quarterly Review of Biology* 46 (1971): 35–57.

12. Pinker, *How the Mind Works*, 404.

13. Pinker, *How the Mind Works*, 404–405.

14. As Peter Railton notes ("The Affective Dog and Its Rational Tale: Intuition and Attunement," *Ethics* 124 (4) (2014): 813–859), we can accept that the moral emotions play the critical functional roles discussed here without accepting many of the other widely held commitments of psychologists, e.g., that moral emotions are fast, automatic, button-pushing responses of the affective system. Only the critical functional roles matter for the purposes of my discussion here.

15. Antoine Bechara, et. al., "Insensitivity to Future Consequences Following Damage to Human Prefrontal Cortex," *Cognition* 50 (1994): 7–8.

16. Bechara, et. al., "Insensitivity to Future Consequences," 14.

17. Bechara, et. al., "Insensitivity to Future Consequences," 14. I have no objection to Bechara and Damasio's use of "knowledge" here. For those who do, replace "knowledge" with "information" or "representation." This will not affect the point.

18. Antonio Damasio, *Descartes' Error* (New York: Grosset/Putnam, 1994).

19. Pardo and Patterson, *Minds, Brains, and Law*, 40–41.

20. Pardo and Patterson, *Minds, Brains, and Law*, 41.

21. I'll return to this point at the end of the chapter.

22. Pardo and Patterson, *Minds, Brains, and Law*, 41 (footnote 92).

23. William Lycan, "Homuncular Functionalism Meets PDP," in *Philosophy and Connectionist Theory*, ed. William Ramsey, et. al. (Hillsdale, NJ: Erlbaum, 1991), 279.

24. Lycan, "Homuncular Functionalism," 277.

25. Lycan, "Homuncular Functionalism," 277.

26. Lycan, "Homuncular Functionalism," 279.

27. Andy Clark, *Mindware* (New York: Oxford University Press, 2014), 60.

28. Alexander Rosenberg, *Philosophy of Social Science* (Boulder, CO: Westview Press, 2012), 39. As stated, [L] is a clear nonstarter. To make it at all plausible, we need to add that there is no obvious better means than *a* available, *a* has to be reasonably low risk compared to the benefit of securing *d*, and so on. While there are serious doubts about whether we can fill out [L] to make it both plausible and nontrivial, the issue won't matter to the argument here.

29. Jerry Fodor, "A Modal Argument for Narrow Content," *Journal of Philosophy* 88 (1991): 19.

30. The Hempel-Hume account of explanation is developed further in Martin Roth, "Folk Psychology as Science," *Synthese* 190 (17) (2013): 3971–3982.

31. Terence Horgan and James Woodward, "Folk Psychology is Here to Stay," *Philosophical Review* 94 (2) (1985): 197.

32. Fodor, *Psychosemantics*, 13.

33. Robert Cummins, "Neo-teleology," in *Functions: New Essays in the Philosophy of Psychology and Biology*, ed. Ariew, et. al. (New York: Oxford University Press, 2002), 172.

34. Rupert, "The Problem of Metaphysically Necessary Effects," 258.

35. Bas van Fraassen, *Laws and Symmetry* (Oxford: Oxford University Press, 1989);Nancy Cartwright, *The Dappled World* (Cambridge: Cambridge University Press, 1999); Paul Teller, "How We Dapple the World," *Philosophy of Science* 71 (2004): 425–447.

36. Ronald Giere, *Explaining Science* (Chicago: University of Chicago Press, 1988).

37. Karl Popper, *Objective Knowledge* (Oxford: Oxford University Press, 1972); William Newton-Smith, *The Rationality of Science* (London: Routledge & Kegan Paul, 1981).

38. David Miller, "Popper's Qualitative Theory of Verisimilitude," *British Journal for the Philosophy of Science* 25 (1974): 166–177; Larry Laudan, "A Confutation of Convergent Realism, *Philosophy of Science* 48 (1981): 19–48.

39. Heidi Maibom ("The Mindreader as Scientist," *Mind and Language* 18 (2003): 296–315) and Peter Godfrey-Smith ("Folk Psychology as a Model," *Philosopher's Imprint* 5 (2005): 1–16) have appealed to Giere's account of models to resolve certain disputes concerning what the folk are doing, psychologically speaking, when they predict and explain the actions of others. According to both Maibom and Godfrey-Smith, our quotidian capacities to predict and explain human action are best understood in terms of facility with a model. However, the appeal to models in this chapter has an importantly different aim: Maibom and Godfrey-Smith are concerned with the psychology of mental state attribution, not whether principles like [L] are contingent or conceptual, or the possible role [L] might play in the science of psychology. Thus, while the arguments presented here may complement the views of Maibom and Godfrey-Smith, it is not an extension of their project. For an appeal to models whose aim is closer to mine, see Menzies, "Reasons and Causes Revisited."

40. Ronald Giere, *Science without Laws* (Chicago: University of Chicago Press, 1999).

41. Paul Teller, "Provisional Knowledge," in *Constituting Objectivity*, ed. Michel Bitbol, et. al. (Netherlands: Springer, 2009), 506.

42. Giere, *Explaining Science*.

43. Horgan and Woodward, "Folk Psychology," 223.

44. Godfrey-Smith, "Folk Psychology as Model," 13.

45. Just as it will be true by definition that Newton's laws apply to any system that satisfies a Newtonian model. In both cases we are assuming that the fit between model and system is exact. While this assumption is implausible, it is not relevant to the present argument.

46. Of course, if some system implements a design, then it will be true by definition that there are components that satisfy these characterizations. This is just what implementing a design comes to. But of any *particular* component, it won't be true by definition that *it* satisfies the functional characterization of belief or of desire (and so forth).

47. Paul Churchland, "Eliminative Materialism and the Propositional Attitudes," *Journal of Philosophy* 78 (2) (1981): 67–90.

48. Pardo and Patterson, *Minds, Brains, and* Law, 41.

49. Pardo and Patterson, *Minds, Brains, and Law*, 29. Pardo and Patterson characterize eliminative materialism as a form of reductionism that seeks the "full-scale elimination of folk psychology" (31), and while their criticism focuses more on the work of Patricia Churchland, they include Paul Churchland as among those advocating this form of reductionism.

Chapter 7

Bad Brains

In the August 4, 1999, edition of *The Onion*, there appeared the following article:

> RITALIN CURES NEXT PICASSO
>
> WORCESTER, MA—Area 7-year-old Douglas Castellano's unbridled energy and creativity are no longer a problem thanks to Ritalin, doctors for the child announced Monday. "After years of failed attempts to stop Douglas' uncontrollable bouts of self-expression, we have finally found success with Ritalin," Dr. Irwin Schraeger said. "For the first time in his life, Douglas can actually sit down and not think about lots of things at once." Castellano's parents reported that the cured child no longer tries to draw on everything in sight, calming down enough to show an interest in television. (Reprinted with the permission of The Onion. Copyright © 2016, by Onion, Inc. www.theonion.com).

This article is obviously satirical (as articles in *The Onion* are wont to be), but what is the target of the satire, exactly? Maybe the article is casting doubt over the reality of conditions like Attention Deficit Disorder (ADD) and Attention Deficit Hyperactivity Disorder (ADHD), conditions Ritalin is often used to treat. After all, the title of the article suggests that Douglas's behavioral dispositions are the dispositions required to be a great artist, and because we value great artists, Douglas does not stand in need of curing. There may be a problem here—perhaps with schools or parents—but it isn't with kids like Douglas. This article will no doubt appear ignorant and misleading to those who think ADD and ADHD are real conditions, but there is a subtler reading of the article available, one that complicates the issue of whether ADD and ADHD are real conditions. On this alternate reading, the article is drawing our attention to the fact that attributions of disease, defect, or malfunction are always *relational*.

FUNCTIONAL ANALYSIS AND NORMS

To ascribe a function to something is to do more than say what the thing does; it is to say what a thing is *for* and, thus, what it is *supposed to do* or what it *ought to do*. Any account of functions must accommodate this normative dimension of functions, and functional analysis does so by relativizing failures to function properly to a target capacity:

> Component *x* is failing to function properly, relative to a capacity *C* of the containing system *S*, if (other things equal) *S* fails to exhibit *C* (or *C* is disrupted) because of what *x* is doing.[1]

This kind of normativity is instrumental, or end-relational,[2] in that it accounts for what something *ought* to do in terms of what something *needs* to do in order for the containing system to exhibit a target capacity. Though relational, the norms functional analysis generates are perfectly objective. Being objective is not the same thing as being intrinsic, as the example of motion makes clear. Thus, while the truth conditions for, e.g., 'the heart ought to pump blood' relativize pumping blood to a capacity of some containing system, they do not relativize it to preferences or beliefs about blood pumping.[3] For example, it can be true that Lassie's heart ought to pump blood regardless of whether she, her owners, or any other observers have a clue about hearts or blood pumping.

We *do* have a clue about many norms, of course, and sensitivity to norms is ubiquitous in living creatures. Consider the regulation of body temperature. For humans, body temperature needs to be around 98.6° Fahrenheit in order to survive, and thus, relative to survival, human body temperature *ought* to be around 98.6°. Notably, human body temperature can be maintained around 98.6° despite fluctuations in external temperature. The explanation of such homeostasis is that the hypothalamus, a complex neural circuit, detects when body temperature strays too far from 98.6° and initiates a cascade of reactions that tends either to decrease or increase heat (depending on the case). For example, if body temperature strays too far below 98.6°, neural activity in the hypothalamus will induce involuntarily muscle contraction (shivering). This muscle contraction generates heat, typically resulting in an increase in body temperature. Relative to regulating temperature, the hypothalamus needs to detect changes in temperature and initiate these cascades. This licenses the claim that the *function* of the hypothalamus is to detect strays from 98.6° and initiate these cascades, and thus that the hypothalamus *ought* to do those things. And because human body temperature ought to be around 98.6°, detecting strays from that temperature *is* detecting that the temperature is not as it ought to be. Furthermore, since the design of the system is such that the cascade of reactions that tends to lower or raise body temperature is initiated by detecting that the temperature is not as it ought to be, there is a perfectly straightforward sense in which the fact that body temperature ought to be 98.6° explains why body temperature of 98.6° is maintained (ceteris paribus). Finally, since the function of the hypothalamus is to detect strays from 98.6° (again, relative to regulating human body temperature), we can make sense of the *mis*detection of norms, e.g., neurons that are supposed to fire only when body temperature is not as it ought to fire when body temperature is as it ought to be. In this way, we can account for the fact that perceptual states can be appraised as correct or incorrect.

At this point it may be objected that *body temperature not being as it ought to be* does not cause anything to happen; rather, all the causal work is done by *body temperature straying from 98.6°*. The objection rests on the thought that if *body temperature not being as it ought to be* does not reduce to *body temperature straying from 98.6°*, then there are two candidate causal properties here. Because the latter realizes the former, the latter has a kind of ontological priority that makes it the better candidate. But then it looks like the nonnormative realizer property—*body temperature straying from 98.6°*—excludes *body temperature not being as it ought to be* from doing any causal work.[4] Denying reduction renders the normative property idle.

We met with a similar objection when we were considering whether reasons can be causes, and the reply given there applies here. Suppose we concede that, in the human case, the property *body temperature not being as it ought to be* is realized by the property *body temperature straying from 98.6°*, i.e., suppose we grant the "local reduction" of *body temperature not being as it ought to be* to *body temperature straying from 98.6*. For those who think normative explanations are causal explanations of events, granting local reduction will rightly appear to undermine any special benefit the explanation couched in normative terms is thought to provide. But the normative explanation at work here inherits its explanatory power from functional analysis, and as we have seen, the explanatory power of functional analysis is not found in causal explanations of events. It is found, rather, in design explanations of capacities. The normative explanation, like functional explanation generally, allows common design to emerge from a welter of noisy detail, and it is this causal relevance filtering that allows us to understand how complex systems work and how they can be modified—and thus evolve. For example, thermal homeostasis is common to all endotherms, and the mechanisms that ensure this are all basically the same. There is a common design here, but dogs hover at a different temperature than humans. A design specification given in terms of a norm—a parameter that takes different values for different creatures—allows us to see a commonality that would be missed otherwise.

The relativization functional analysis delivers is just what Carl Craver calls perspectivalism; you get an ought relative to a particular design and the capacity such a design explains.[5] The design isn't relative, of course; if the system has it, that is an objective fact, as are the norms that come with that design. For example, a temperature of 98.6° isn't the temperature you ought to have if you have an infection. Bacteria are temperature sensitive, and the higher the temperature, the more iron they require for metabolizing. Vertebrate immune systems raise temperature and lower iron in response to bacteria; relative to fighting infections, this is what immune systems need to do and thus ought to do. Norms are thus variable across species and conditions. Moreover, they come with trade-offs: optimal flight isn't optimal metabolism, and optimal temperature isn't optimal immune response. These trade-offs are inevitable whenever you embed a subsystem in a complex design. For these reasons, we should be skeptical that nonrelational norms have a legitimate role to play in science or in any other venue. A relational norm may appear to be a nonrelational norm if we privilege a particular goal-state, but the relational normativity is still there.[6]

So is there something wrong with Douglas or not? This is a bad question. Let's start with the claim that there is something wrong with his behavior, i.e., he is not behaving as he *ought* to behave. How ought he to behave? Consider his behavior in the context of the modern public school system, a system in which teachers often face overcrowded classrooms. It is plausible that learning can occur in such a setting only if each student is quiet and remains in his seat; relative to the school system's capacity to teach children, Douglas ought to sit quietly because he needs to sit quietly in order for this capacity to be exercised. However, suppose we consider Douglas's behavioral dispositions in light of society's capacity to maintain a thriving art world. Relative to that capacity, we might say that Douglas has the dispositions he ought to have, for these are the dispositions (at least some) members need to have in order to maintain a thriving art world. With this point in mind, let's now consider the suggestion that there is something wrong with Douglas's brain, something that is "cured" by Ritalin. It should be fairly obvious that our assessment of Douglas's brain is parasitic on our assessment of his behavior, i.e., if we think that Douglas has the behavioral dispositions he ought to have, we will have no grounds for thinking that there is something wrong with his brain. However, even if we set aside the issue of normative relativity, the inference from behavior to brain is not direct. If the norms relative to which we are attributing disease, defect, or malfunction are the norms of the modern public school system, failure to conform behavior to these norms does not automatically imply that there is something defective about the person who fails to conform. Although children may have an especially hard time conforming their behavior to norms, we don't completely abandon the participant perspective toward them, and to the extent that we hold a child responsible for her failures to conform, to that extent we don't view the child as someone who needs *curing*. Put another way, from the fact that a person does something wrong, it does not follow that there is something wrong with the person. In the present context, our concern with whether there is something wrong with the person is a concern about whether the person possesses the capacities necessary to make adopting the participant perspective toward that person reasonable. If the person does possess such capacities, then although the person may fail to conform her behavior to the relevant norms, we believe that conformance can be achieved from within the participant perspective, i.e., we do not view the person as defective qua moral agent.

Our judgment that there is something wrong with Douglas's brain thus rests on three assumptions: (i) there is something wrong with Douglas's behavior; (ii) the brain causes behavior; (iii) conformance to the norm cannot be achieved within the participant perspective.[7] What do these assumptions imply about the "reality" of neurological disorders like ADD and ADHD? One implication is that if such disorders are real, their reality qua disorders won't be an intrinsic property of the brain. This simply follows from (i), and as we've seen, it is plausible to think that these conditions are considered disorders precisely because of the way these conditions impact behavior. But when we keep in mind that the normative assessment of behavior is itself relational, e.g., (i) is true relative to the norms of modern public schools but not true relative to the norms of artistic achievement, then the very same brain can be both disordered and not disordered. This is no more contradictory than the claim that a body is both moving and not moving; because claims about motion implicitly refer to a frame of reference, it can be true both that a body is moving with respect to one frame of reference and not moving with respect to another frame of reference. One reply here might be to argue that some contexts are more significant than others, and the more significant a context, the more normative weight it carries.[8] For example, if we could defend the idea that learning is more significant than artistic achievement, then we might have a plausible basis for giving priority to the assessment that Douglas is not behaving as he ought to. While such a defense might be made, the point here is simply to draw our attention to the fact that there is more than one context available for normative assessment. Notice, too, that this point applies even if we assume that learning is more significant than artistic achievement. Our assessment that there is something wrong with Douglas's behavior is based on the norms of behavior found in modern public schools. But those norms are not the norms needed for learning; they are the norms needed for learning given the current realities of modern public schools. Those realities reflect our political and economic priorities, and those priorities can be revised. Imagine a school system in which each student received individualized instruction tailored to the temperament and interests of the child. Would we think there is a problem with Douglas's behavior? Plausibly, we would not.[9]

Given the facts about Douglas's brain, he is unable to conform his behavior to the norms of modern public schools. The question now is what we should do about it. One response is to hold the norms fixed and alter Douglas's brain in a way that enables him to conform: the problem is with Douglas's brain. Another response is to leave Douglas's brain alone and alter the norms: the problem is with the norms. The point of my argument is not that one response is better than the other, but rather to show that there really are two responses available. Once we recognize this, we actually have to do the hard work of defending our responses. Part of what makes defending the first response hard is that we cannot simply declare that Douglas has a neurological disorder. That declaration isn't a *defense* of the first response; it *is* the first response. But enough about Douglas. It is now time to consider the implications of all of this for the law.

CONFORMANCE TO LAW AS NORM

In the context of our legal system, the relevant norms of behavior are provided by the laws themselves: you ought to obey the law. As discussed in Chapter 3, oughts imply cans, so if you ought to obey the law, then you can obey the law. However, we saw that the legitimacy of a demand, e.g., a law, also turns on the content of the demand. A law is not legitimate simply because we enact the law and people can obey it. In protest to slavery, Henry David Thoreau refused to pay a poll tax and went to jail. According to one version of the story, Ralph Waldo Emerson visited Thoreau and asked him why he was in jail, and Thoreau responded by asking Emerson why he *wasn't* in jail.[10] Unjust laws ought to be broken, not obeyed. We can understand the general point even if we don't agree with Thoreau's actions in particular, a point that has been appreciated at least since Plato: laws are not legitimate because we enact them; we enact laws because they are legitimate.

My primary aim in raising the issue of legitimacy here is not to bring the legitimacy of our laws into question, but to make explicit the crucial role that the assumption of legitimacy plays in our legal system. A central theme of this book is that we can locate legal responsibility within a broader conception of responsibility, a conception of responsibility that is essentially linked to rationality. The law operates with the default policy that people who break the law are legally responsible for their crimes. This policy isn't arbitrary nor are the exceptions we make to it. If we hold a person legally responsible for her crime, then our reason for holding her responsible should also provide her a reason to take responsibility. In taking responsibility for her crime, the person recognizes that she did not meet a demand that she was able to meet. This is not enough, however. In order for her failure to have normative significance, the person must also view the demand as legitimate. If taking legal responsibility is a specific instance of taking responsibility for failing to meet rational demands, then it will make sense to take legal responsibility for breaking the law only if it is rational to obey the law.

Under what conditions is it rational to obey the law? Suppose you know that you will be severely punished if you break any law, regardless of whether you view the law as legitimate. Given that you don't want to be severely punished, and given that you believe you can avoid severe punishment by obeying the law, this is a situation in which it would be rational to obey the law. Clearly this is not the sense of rationality that we're after, and we can rule this kind of case out by directing our attention to the content of a law: under what conditions is the content of a law such that it is rational to obey it? The law is primarily concerned with the sphere of action, and with respect to practical reason, a D-rational agent is an agent who knows what is good and wills what is good. In Chapter 4 I left open the question of what kinds of things count as good, but I also said it is plausible that the things commonly identified as good—meaningful work, close friendships, mental and physical health, and so forth—are what we are after. For present purposes, however, it is enough that we accept that there is an important sense in which it is rational to obey laws and that the rationality of obeying laws has to do with the content of laws. I don't think this will be terribly controversial, and again, my point in mentioning it is mostly to note the role this assumption plays in explaining the normative significance of failure for the person who commits a crime, a significance without which we cannot make sense of the practice of taking responsibility—and all that comes with it, including censure and punishment.

In Chapter 2 we met with the idea that, as far as responsibility goes, all the law cares about is whether you are rational. On one reading of this claim, what the law cares about is whether you knew what you were doing—you weren't delusional, for example. But the discussion in the previous paragraph suggests another way of articulating the concern about rationality: if it is rational to obey the law, then people who break the law are acting irrationally. Of course, acting irrationally is not the same as being irrational, in the sense of being irrational that is supposed to undermine responsibility (e.g., insanity). When a person acts irrationally, we want to know whether the person is A-rational. An A-rational person will acknowledge the normative significance of breaking the law, and because it was *her* failure, she will accept that the censure and punishment she receives is merited. Being A-rational, she will also see herself as having a reason to take steps that will help bring her future behavior into compliance with the law. Short of no crime at all, the ideal situation is one in which criminals respond to their crimes in just the way Spiderman responded to Ben's death. The task of the next section is to consider the conditions that would prevent this ideal from being achieved, and what implications this would have for law. Before doing so, however, I want to say a bit more about how the account of norms delivered by functional analysis fits into the overall aim of fusing the manifest and scientific images.

In the previous section I talked about the way "ought to" tracks "needs to," and so if it is true that we ought to follow the law, then following the law is what we need to do if some capacity is to be exercised. In a related vein, in Chapter 3 I said the function of the CAA is to produce D-responsible people; the CAA *ought* to produce D-responsible people because this is what the CAA *needs* to do in order for a containing system to exhibit a target capacity. What is the containing system here? What is the target capacity? How does the capacity depend on D-responsible agents? The following remark by Daniel Dennett suggests the beginnings of an answer:

> The *consequences* of the policy of holding one's self and others responsible justify the policy. Why? Because the mutual presumptions of competence that this entails support a host of societal activities: trade, promising, multi-agent/multi-year projects. Civilization, in effect, depends on the practice of holding ourselves and other responsible.[11]

Dennett talks of justification, but we can put the point in terms of explanation, namely, an explanation that links these practices to the implied benefits of civilization. Civilization depends on the policy of holding one's self and others responsible because civilization is supported by competences entailed by the policy. Which competences are those? The activities Dennett mentions all involve some form of cooperation, and investigation into the psychology of cooperation provides powerful evidence that human beings are reciprocal altruists.[12] In short, we tend to help others if and only if those others help us. The basic rationale for the widespread existence of this policy is that cooperation tends to make each person better off than they would be otherwise. A simple example illustrates the point. Shelter protects me from an often hostile environment, but I cannot build a house by myself. You are in the same situation. But *together* we can build a house, so if we cooperate each of us ends up with shelter. As work in game theory makes clear, however, there is a serious fly in the ointment.[13] You are the only one who benefits when I help you build your house, and helping you actually comes at a cost to me. While this may be a cost I am willing to bear if you help me build my house, I would actually be better off if you helped me build my house but I didn't help you build your house. So suppose I ask you to help me build my house, promising to help you build your house when we are done. You realize that I would be better off not helping you after you help me, so you know that I'd have a good reason to break my promise. But you are not going to help me if I don't help you, so you conclude that you won't help me. Of course, I'll reason the same way if you ask me to help build your house first. The result is no one gets a house built, even though we recognize that each of us would be better off with a house.

But what if we could somehow create a situation in which promises made are promises kept, i.e., what if your act of making a promise to help me build my house was a reliable indicator of future cooperation on your part? In such a case I would be willing to help you build your house and each of us would benefit. To derive the benefits of cooperation, then, we need creatures who have the *right to make promises*—in short, creatures who are D-responsible. There are no perfectly D-responsible agents, of course, and sometimes we give into the temptation to cheat (not reciprocate). Because civilization depends on D-responsible agents, or *nearly* D-responsible agents, there needs to be some way to prevent cheating from becoming rampant and destroying cooperation altogether. The practice of holding one's self and others responsible is a means by which this is done, and a means by which (nearly) D-responsible agents are created. A-responsibility is needed for this practice to work, and it is in this sense that the function of the CAA is to produce D-responsible agents. Moreover, the CAA can serve this function only if it has the right sort of design, namely, a design that enables A-responsibility.

The project of showing how ethical norms and practices can be located in the natural order of things has been with us at least since Hobbes, and while it is beyond the scope of this book to show that the account of norms provided by functional analysis can get us all the way to ethics (and law), it is worth providing a rough outline of how it might go. Properties like being morally right (permissible, obligatory) are evidently multiply realized—i.e., the physiology, and even the psychology, of all morally right actions share nothing of interest in common beyond being morally right. This, by itself, is enough to defeat reductionism in metaethics, just as it is enough to defeat reductionism in the philosophy of mind. Yet a major advantage of functionalism in the philosophy of mind is precisely that it is a naturalistic theory. Thus, just as nonreductive functionalism holds that every particular mental process is some physical process or other, so an analogous nonreductive naturalistic metaethics would hold that every morally right action is some physical process or other. And just as nonreductive functionalism holds that a particular realizer of a mental process is a mental process, not because of its intrinsic physical properties, but because of its role in a complex system, so a nonreductive naturalistic metaethics would hold that a particular realizer of a morally right action is a morally right action, not because of its intrinsic physical properties, but because of its role in a complex system of social interaction.[14] While this will not generate *the* norm—social systems are as messy and complex as biological systems—it is also plain that the relativity in ethics isn't a failure of objectivity, i.e., a person can be subject to norms that the person doesn't acknowledge or accept. In the case of health, crank addicts demonstrate that flouting what is normal fractures your body and mind. In the case of ethics, psychopaths demonstrate that flouting what is normal fractures relations of reciprocal altruism.

Naturalists insisted on reduction, and their critics denied its possibility. Ethical realists therefore looked like dualists. This is exactly what happened in the philosophy of mind until functionalism changed the debate. No one is a dualist about doorstops. Moreover, it became clear that failures of reduction have no metaphysical significance no matter where they occur. The scientific, naturalistic worldview has to be satisfied with supervenience. That said, who wants to deny that the norms of human behavior—human values and obligations—supervene on human biology and psychology? The real question is to understand that supervenience, to understand how it is possible.

What makes naturalism in ethics appealing is the thought that a viable epistemology for ethics will come along for the ride: a naturalistic ethics would be an ethics for which we could tell a plausible story about how it is that values and norms are known. A naturalist, it seems, can reasonably hold that it is possible, in principle at least, to establish by observation and experiment that is it wrong to stone rape victims just as it eventually proved possible to establish that the earth moves and is not at the center of the universe, and that living things evolve by natural selection. These were hard won achievements and are still a long way from universal acceptance. Furthermore, the underlying epistemology is, in its details, still a matter of considerable debate. But when we think about these issues through the lens of functional analysis, the prospect of naturalizing objective values looks much brighter.

DEFECTIVE BRAINS AND EXISTING LAW

The account of free will developed in this book implies that there can be degrees of free will, that is, variability in the extent to which someone can respond in the way distinctive of free persons. This variability is reflected in the variability of norms we apply to behavior, including age restrictions for driving, voting, and drinking, and in a host of other formal and informal ways. But we do largely assume that the members of this community are responsive. I argue that this assumption provides the principle rationale for judgments of criminal responsibility, at least in what we may think of as the ideal case. Two conditions are satisfied in the ideal case: (1) the person judged to be criminally responsible recognizes that her actions deviated from norms of behavior that she has good reasons to endorse; (2) the person judged to be criminally responsible has the capacity to bring her actions into compliance with these norms. If both conditions are met, we should reasonably expect the person to seek forgiveness, i.e., to acknowledge that the attitude displayed in her actions was such as might properly be resented and to repudiate that attitude for the future.[15] The legal significance of seeking forgiveness is reflected in the practice of parole, and so another way to put the point is that, in the ideal case, every person convicted of a crime would satisfy the conditions of parole.[16]

I've argued that the true threat to free will is not determinism, but rather various forms of ignorance or failures to coordinate understanding and will. This may seem to imply that any occasion in which a person acts from ignorance or fails to coordinate understanding and will is an occasion in which the person lacks free will. If free will is a precondition for responsibility, and if every crime results from either ignorance or a failure to coordinate understanding and will, then no one will be responsible for committing a crime. But as I noted in Chapter 4, these failures of understanding and will do not by themselves undermine free will, for what is crucial to having free will is the capacity to respond appropriately to such failures. This appropriateness was characterized in terms of an ideal of rationality that can play a regulatory role in thought and behavior. Those for whom the ideal does play this regulatory role I called A-rational, and I said that whether a person has free will turns principally on whether the person is A-rational. A consequence of this claim is that conditions which threaten A-rationality are conditions which threaten responsibility. Because failures of rationality come in two forms—failures of understanding (e.g., ignorance) and failures to coordinate understanding and will—conditions that threaten responsibility are conditions that compromise a person's capacity to respond appropriately to one or both kinds of failure. Legal theory and practice provides independent support for this position, as I will now show.

Let's start with ignorance. Suppose I am walking down a crowded street and notice a gun on the ground. I pick up the gun, point it at a passer-by, and pull the trigger. The gun fires and kills the passer-by. I may not have known the gun was loaded—I acted in ignorance of that fact—but such ignorance does not relieve me of criminal responsibility for the death. But why not? I recognize the legitimacy of the demand not to kill innocents, but given my ignorance, perhaps it was not within my power to conform my behavior to this demand. Here we lean on counterfactuals about what I would have done had circumstances been different, and one of the relevant counterfactuals concerns what I would have done had I known the gun was loaded. Let's stipulate that I would not have pulled the trigger if I had known the gun was loaded. I didn't know the gun was loaded, of course, but now we can ask whether there was a legitimate demand on me to know whether the gun was loaded. If I recognize the legitimacy of the demand not to kill innocents and recognize that whether I meet that demand depends on whether the gun is loaded, then I will see myself as having a good reason to meet the knowledge demand. Whether it is within my power to meet the demand will depend on whether I would have met the demand had circumstances been different, e.g., if I would have known the gun was loaded had I checked. Was it within my power to check? Yes, for I would have checked had I thought to do so. This doesn't set off the kind of regress we worried about in Chapter 2, because our present concern is only whether my ignorance excuses me from criminal responsibility for the death. To answer that question, we need only consider whether certain demands on knowledge are legitimate, i.e., we need only consider whether conditions (1) and (2) are satisfied when applied to the knowledge demand. They are, so I am responsible for my ignorance and thus responsible for the death.

Now suppose a different situation. I am hosting a dinner party, and one of my guests asks for a glass of water. I honor the request, and my guest drinks up. Almost immediately, the guest collapses on the floor and dies. It turns out that in the middle of the previous night an evil mastermind laced my drinking supply with strychnine. I had no good reason to believe anyone was out to harm me, and the mastermind was so skilled as to not leave any noticeable traces of his presence. And because I had not yet tried the water that day (if I had, I wouldn't have been around to host the party), I had no reason for thinking the water was poisoned. Thus I, too, acted in ignorance, but it is plausible to suppose that I would be relieved of criminal responsibility for my guest's death. As far the issue of responsibility goes, the difference between the cases appears to turn on the extent to which the ignorance was justified. In the first case I lacked adequate justification for thinking the gun was not loaded, so either I ought not to have acted as I did or I ought to have taken steps to acquire adequate justification. These oughts imply cans, of course, and it was within my power to do each. In the second case, by contrast, I possessed adequate justification for thinking that the water was not poisoned, and so even if it had been within my power to check, we could not have legitimately demanded that I do so. Put bluntly, the difference in the cases turns on whether *I ought to have known better*. In the first case, the two conditions on the ideal case of criminality are satisfied because they are entailed by the fact that the two conditions are satisfied by my ignorance. In the second case, though it may have been within my power to check, I had very good reasons for thinking the water wasn't poisoned. My ignorance was justified in a way that it wasn't in the first scenario.

The relevance of ignorance to responsibility shows up most clearly in cases involving insanity. In our legal system, the oldest, most widely used "test" for insanity is some version or other of the M'Naghten Rule. On January 24, 1843, Daniel M'Naghten attempted to kill British Prime Minister Robert Peel. Though he failed to kill Peel, M'Naghten did kill Peel's secretary, Edward Drummond. M'Naghten was tried for murder, but was acquitted by reason of insanity. Public outrage over the verdict prompted changes in the British insanity defense, yielding what we now call the M'Naghten Rule:

> [E]very man is to be presumed to be sane … [T]o establish a defense on the ground of insanity, it must be clearly proved, that, at the time of the committing of the act, the party accused was laboring under such a defect of reason, from disease of the mind, as not to know the nature and quality of the act he was doing, or, if he did know it, that he did not know he was doing what was wrong.[17]

As an illustration of the rule, consider how insanity is defined by the State of Illinois:

> (a) A person is not criminally responsible for conduct if at the time of such conduct, as a result of mental disease or mental defect, he lacks substantial capacity to appreciate the criminality of his conduct. (b) The terms "mental disease or mental defect" do not include an abnormality manifested only by repeated criminal or otherwise antisocial conduct.[18]

Insanity is a form of ignorance in that an insane person does not appreciate the criminality of his conduct, i.e., he doesn't *know* that he is committing a crime. It is fairly evident why insanity would thus prevent condition (1) from being satisfied: if a person does not appreciate the criminality of his conduct, he will not recognize that his actions deviate from norms of behavior that he has good reasons to endorse. But to be insane, it is not enough that the person does not appreciate the criminality of his conduct; he must lack *the capacity* to appreciate the criminality of his conduct. Because an insane person lacks this capacity, he cannot know better, and because he cannot know better, it cannot be properly said of him that he ought to have known better. This seems clear enough; a bit more difficult to parse, however, is the role that mental disease or defect is playing in the definition given by (a) and in the qualification provided by (b). Let's start with the definition. It is not enough that a person lacks the capacity to appreciate the criminality of his conduct; the lack of capacity has to be caused by mental disease or defect. There are a couple of ways we might read this. Perhaps the reference to disease or defect is intended to clarify what is meant by "lacks substantial capacity" and to reinforce the idea that insanity requires more than the fact that the person *didn't* appreciate the criminality of his conduct; it also requires that he *couldn't*.

Recalling our earlier discussion of Douglas, it wasn't merely that Douglas didn't conform that licensed our judgment of disorder, but rather that he *couldn't* conform. The reasoning might be: the person didn't appreciate the criminality of his act. If this failure was the result of mental disease or defect, then this would also show that he couldn't appreciate the criminality of his act. This reading of (a) provides a complementary reading of (b): a repeat criminal may also not appreciate the criminality of his acts at the time he performs them, but this does not imply that the repeat criminal couldn't have appreciated the criminality of those acts at the time he performed them. As such, repeat criminality is not sufficient for lacking the capacity to appreciate criminality. On this reading, then, all that matters to being insane is that one lacks the capacity, and the reference to disease or defect is simply a way to underscore what is meant by "lacks substantial capacity."

Alternatively, we might read (a) as stating that to be insane, it isn't enough that the person couldn't have appreciated the criminality of his act; insanity requires that this lack of capacity be caused by mental disease or defect. This reading of (a), too, provides a complementary reading of (b): mental disease or defect may cause some people to be repeat criminals, but by itself, this does not show that the person lacks substantial capacity to appreciate the criminality of his conduct.[19] However, even if this is the legally correct reading, it is deeply problematic. For one, it is not clear what other causes of a lack of capacity it is supposed to be ruling out. For example, imagine that you are kidnapped by an evil sorcerer. The evil sorcerer casts a spell that renders you incapable of appreciating the criminality of acts. You perform a criminal act, get arrested, and are brought to trial. You plead not guilty by reason of insanity, and during the trial the cause of your lack of capacity is discovered. Are you thus not insane, and thus likely to be found legally responsible? If so, then this definition of insanity is, well, insane. But there is another, equally important problem here, one that we also encountered in our discussion of Douglas. Sorcerers notwithstanding, let's assume that if a person lacks the capacity in question, he lacks it as a consequence of his mental condition. Let's also assume that his mental condition is a consequence of his neural condition ("you are your brain"). Responsibility requires possessing the capacity to appreciate the criminality of conduct, i.e., a person needs this capacity in order to be responsible. To possess this capacity, the person needs a brain that is structured in a certain kind of way. If the person lacks this capacity as a result of the way the person's brain is structured, then the brain isn't structured in the way it ought to be, relative to the capacity to appreciate the criminality of an act. But this is precisely the ground for saying that the person has a neurological disease or defect. It does not make sense to say that the person lacks this capacity but has a properly functioning brain, for whether a brain is functioning properly turns on whether the brain contributes what it needs to in order for the person to possess a particular capacity.

While I think the above points are worth making, I also don't want to overstate their practical significance. I doubt we actually find many cases in which, although everyone is convinced that the defendant lacks the capacity in question, lingering, unanswered questions about the cause prevent finding the defendant innocent by reason of insanity. We want to know whether the defendant lacks the capacity, and the neurological evidence that speaks to disease or defect is used primarily to convince jurors that the defendant actually lacks, or doesn't lack, the capacity. The idea is that we may come to know enough about the specific ways the brain enables various capacities to know that a person with a certain kind of brain structure cannot have the capacity in question. To do this, we don't need some prior determination that the brain in question is diseased or defective. Rather, we arrive at that characterization when we become convinced that the kind of brain we're looking at prevents people from having certain capacities. I'll come back to these points in the next chapter. Meanwhile, we need to confront yet another problematic feature of the definition of insanity, one that does have serious practical consequences.

Clause (a) is a conditional statement, and we can make the logical form explicit by rewording it as follows: *if* at the time of such conduct, as a result of mental disease or mental defect, he lacks substantial capacity to appreciate the criminality of his conduct, *then* a person is not criminally responsible for conduct. The reworded sentence sounds awkward because of the way the pronouns work, but the meaning is clear enough; it basically says that lacking substantial capacity to appreciate the criminality of conduct is *sufficient* for not being criminally responsible. But suppose a person *possesses* substantial capacity to appreciate the criminality of conduct. Does the conjunction of that supposition with (a) imply that the person *is* criminally responsible? As a purely logical point, the answer is no. Schematically, (a) has the form "If P, then not Q" and the supposition has the form "not P." These two claims do not entail Q, however, as the following example demonstrates:

If Frisky is a cat, then Frisky is not a reptile
Frisky is not a cat
Therefore, Frisky is a reptile.

Suppose Frisky is a dog. In that case, though the first two premises are true, the conclusion is false. The argument is invalid. Possessing substantial capacity to appreciate criminality of conduct entails being criminally responsible only if lacking substantial capacity is *necessary* for not being criminally responsible, and while the Illinois law doesn't actually say that lacking substantial capacity is necessary for not being criminally responsible, it is generally understood that way. Why? I've argued that the ideal case of criminal responsibility is one in which conditions (1) and (2) are satisfied. If a person possesses substantial capacity to appreciate criminality of conduct, we might reasonably suppose this is enough to satisfy (1). We might also reasonably suppose this is enough to satisfy (2), but only if we make the following assumption: possessing substantial capacity to appreciate criminality of conduct is sufficient for the capacity to conform behavior to law. Supposing you agree with this, there are a number of ways you might respond. You might accept that criminal responsibility requires satisfying (1) and (2) and also accept the assumption that possessing substantial capacity to appreciate criminality of conduct is sufficient for satisfying both (1) and (2). Or you might accept that criminal responsibility requires satisfying (1) and (2) but reject the assumption that possessing substantial capacity to appreciate criminality of conduct is sufficient for satisfying both (1) and (2). Or you might simply reject condition (2) altogether. Only the first and third options justify inferring criminal responsibility from the fact that a person possesses substantial capacity to appreciate criminality of conduct. Where does the law stand on the matter?

A cursory look at the history of legal theory and practice suggests an uneasiness with the first and third options, i.e., within the law itself we find evidence that criminal responsibility requires satisfying conditions (1) and (2) but that possessing substantial capacity to appreciate criminality of conduct is not sufficient for satisfying both conditions. For example, in the 1954 case *Durham v. United States*, the U.S. Supreme Court ruled that a person should be deemed not criminally responsible if the criminal act resulted from mental disease.[20] If an action results from mental disease, it does not follow that the person who commits the act fails to appreciate the criminality of her act. Thus under this ruling it is possible that a person appreciates the criminality of her act and yet is still not criminally responsible. This test for criminal responsibility proved difficult to implement, however, and it was ultimately rejected in the 1972 case *United States v. Brawner*.[21] The guidelines for criminal responsibility provided by the American Law Institute are another example.[22] Introduced in 1962, these guidelines state a person is not criminally responsible for her act if, as a result of a psychiatric disorder, the person lacked control over her impulse to act. Here, too, the fact that a person lacked control over her impulse to act does not imply that the person did not appreciate the criminality of her conduct, so as with the test introduced by *Durham*, under these guidelines a person may appreciate the criminality of her conduct and yet still be deemed not criminally responsible. While no single standard emerged, by the early 1980s many states had incorporated the capacity to control behavior into the test for criminal responsibility.

Each test we've considered so far constitutes an affirmative defense; that is, if a defendant "passes" the test, the defendant is deemed not criminally responsible for her conduct. But this is not the only way such tests can be relevant to questions of responsibility. Under the plea of diminished capacity, a defendant may offer evidence of her mental condition in relation to her capacity to form the *mens rea*, or intent, required for commission of the charged offense.[23] This plea may be raised by a defendant who is legally sane, and while diminished capacity does not constitute an affirmative defense, it can mitigate murder to manslaughter.[24] A defendant is found to have a diminished capacity if the defendant did not appreciate the nature of her conduct *or* was incapable of conforming her conduct as a result of mental disease or defect.[25] The doctrine of diminished capacity thus combines aspects of the M'Naghten Rule and the American Law Institute guidelines. Perhaps the most notorious example of a successful diminished capacity plea is the case of Dan White, a case that generated significant backlash against the use of this plea.[26] In 1979 White killed San Francisco Mayor George Moscone and City Supervisor Harvey Milk, and in the famous "Twinkie" defense, White's lawyer argued that a junk food diet prevented White from forming the intent necessary for premeditation. White was convicted of voluntary manslaughter, and the resulting public outrage led to the elimination of the diminished capacity plea in a number of states, including California. Affirmative defenses that appealed to the inability to control behavior met with a similar fate. In 1981 John Hinckley attempted to assassinate President Ronald Reagan, and at trial, Hinckley's lawyers successfully argued that Hinckley was unable to control his behavior. Hinckley was found not guilty by reason of insanity, and within a few years most states had reverted to the exclusive use of some version of the M'Naghten Rule.[27]

Nonetheless, doubts about the adequacy of M'Naghten type tests persist. In the 2002 case *Atkins v. Virginia*, the U.S. Supreme Court ruled that the execution of mentally retarded persons violates the Eighth Amendment ban on cruel and unusual punishment. The Supreme Court wrote that

> Mentally retarded persons frequently know the difference between right and wrong and are competent to stand trial. Because of their impairments, however, by definition they have diminished capacities to understand and process information, to communicate, to abstract from mistakes and learn from experience, to engage in logical reasoning, to control impulses, and to understand the reactions of others. There is no evidence that they are more likely to engage in criminal conduct than others, but there is abundant evidence that they often act on impulse rather than pursuant to a premeditated plan, and that in group settings they are followers rather than leaders. Their deficiencies do not warrant an exemption from criminal sanctions, but they do diminish their personal culpability.[28]

Because knowing the difference between right and wrong is more or less the same as appreciating the criminality of an act, a mentally retarded person who knows the difference between right and wrong will likely not be found legally insane under a M'Naghten type test. But as the language of the ruling makes clear, the fact that a person appreciated the criminality of her act is not enough to settle the extent of her responsibility. Moreover, the factors relevant to diminishing culpability are factors relevant to whether one has the capacity to conform behavior to the law. I said that, in the ideal case, a person who is criminally responsible for her actions is a person who would seek forgiveness. But seeking forgiveness for an act involves the pledge to not act that way again, and such pledges are empty if one is unable to learn from mistakes or is unable to alter the effect impulses have on behavior.[29]

In *Atkins*, the question before the Supreme Court was whether the execution of mentally retarded persons violated the prohibition on cruel and unusual punishment. The Court ruled that it did, and we've seen the reasons that were offered in support of that ruling. What makes *Atkins* especially significant, however, is the potentially broader impact of the ruling. In particular, I want to consider the implications of *Atkins* for how we think of addiction and psychopathy. The American Society of Addiction Medicine characterizes addiction as a

> primary, chronic disease of brain reward, motivation, memory and related circuitry. Dysfunction in these circuits leads to characteristic biological, psychological, social and spiritual manifestations. This is reflected in an individual pathologically pursuing reward and/or relief by substance use and other behaviors. Addiction is characterized by inability to consistently abstain, impairment in behavioral control, craving, diminished recognition of significant problems with one's behaviors and interpersonal relationships, and a dysfunctional emotional response.[30]

And here is how *Scientific American* describes psychopathy:

> Superficially charming, psychopaths tend to make a good first impression on others and often strike observers as remarkably normal. Yet they are self-centered, dishonest and undependable, and at times they engage in irresponsible behavior for no apparent reason other than the sheer fun of it. Largely devoid of guilt, empathy and love, they have casual and callous interpersonal and romantic relationships. Psychopaths routinely offer excuses for their reckless and often outrageous actions, placing blame on others instead. They rarely learn from their mistakes or benefit from negative feedback, and they have difficulty inhibiting their impulses.[31]

When we compare these characterizations to each other, and compare each to the characterization of mentally retarded persons provided in *Atkins*, we notice a significant overlap of features. Three features appear common to all:

(i) difficulties with controlling impulses; (ii) difficulties with establishing and maintaining healthy interpersonal relationships; (iii) difficulties with learning from mistakes. There are also important differences, of course. Unlike mental retardation, neither addiction nor psychopathy seems to involve difficulties with information processing or logical reasoning. But like the brain damaged patients studied by Bechara and Damasio, each condition appears to involve an impaired ability to integrate cognitive function with decision-making.[32]

In our earlier discussion of Illinois' definition of insanity, I allowed that repeat criminality and a tendency toward antisocial behavior by itself does not entail that a person lacks the capacity to appreciate the criminality of conduct or the appropriateness of norms. At issue in that discussion was condition (1) of the ideal of responsibility, but might repeat criminality or the tendency toward antisocial behavior be sufficient for failing to satisfy (2)? No. Recall our discussion of the student who routinely fails to do what a student ought to do; this student is irresponsible in the sense of not being a D-responsible student. However, this does not imply the student is incorrigible, and some irresponsible students can become D-responsible students by taking responsibility for their failures. Whether taking responsibility can have this effect depends on whether the person is A-responsible, but even then we cannot assume that all A-responsible people become D-responsible people. To be A-responsible is to be animated by certain demands, and while this animation requires that the demands in question *can* serve as effective reasons for action, it does not mean that they *always* serve as effective reasons. Moreover, even when such demands do serve as effective reasons, there is no guarantee that the person's action will meet with success. Consider the *Atkins* ruling. We might be tempted to think that a mentally retarded person can satisfy (1) but not (2), and that is why such persons are not fully responsible. But the ruling did not say mentally retarded persons cannot satisfy (2); it said mentally retarded persons have impairments that compromise their ability to satisfy (2). *That* is why they are not fully legally responsible. Like the A-responsibility on which it depends, legal responsibility is a matter of degree.

With these points in mind, let's consider addiction and psychopathy. In Chapter 3 we noted that while Strawson did not think determinism per se ruled out responsibility, he did acknowledge situations that justify abandoning the participant perspective toward others. For example, we might come to abandon the participant perspective toward an agent if

> we see the agent … as one whose picture of the world is an insane delusion; or as one whose behavior, or a part of whose behavior, is unintelligible to us, perhaps even to him, in terms of conscious purposes, and intelligible only in terms of unconscious purposes: or even, perhaps, as one wholly impervious to the self-reactive attitudes I spoke of, whole lacking, as we say, in moral sense.[33]

The characterizations Strawson provides are not meant to be empirical descriptions of actual people, e.g., perhaps no one is *wholly* impervious to the self-reactive attitudes. But these simplifications can be useful to identifying what is relevant to moral responsibility, and why. This passage draws our attention to three of those features. The first is evidently related to our earlier discussion of insanity, but I want to focus on the second two features and the way they can help us think about the implications of addiction and psychopathy for responsibility. Let's start with psychopathy, exemplified by the person who is wholly impervious to the self-reactive attitudes. Such a person may understand the content of certain demands and recognize that his behavior violates these demands, but the recognition of such violations does not invoke the appropriate self-reactive attitudes, e.g., remorse, shame, and so forth. If appreciating the normative force of a demand consists in having the appropriate self-reactive attitudes, then psychopaths cannot appreciate the normative force of demands.[34] Given the role that self-reactive attitudes play in bringing about changes in one's CAA, demands made of such people cannot be transformed into effective reasons. Now consider addiction. A person who suffers from addiction is much like the person whose behavior is unintelligible in terms of conscious purposes. An addict may understand the content of demands and recognize that his behavior violates such demands, and the recognition of such violations may very well invoke the appropriate self-reactive attitudes. Unlike the psychopath, the violations can *matter* to the addict. But while such violations may matter to the addict, because of the addiction the self-reactive attitudes do not bring about the kinds of changes in the CAA that transform demands into effective reasons for action. As a result, the addict may routinely find himself in situations in which he is conscious of behaving in a way that conflicts with how he thinks he ought to act—he sees himself as having a compelling reason not to act the way he does—and it is in this sense that the addict's behavior is unintelligible to himself. The crucial difference between the psychopath and the addict is that the latter, but not the former, can appreciate the normative force of certain demands. Such appreciation by the addict explains why his violations render his behavior unintelligible. By contrast, in failing to appreciate the normative force of demands, the psychopath's behavior does not present itself as conflicting with what he thinks he has a compelling reason to do. Note that this may help explain why psychopaths blame others for behavior; because the psychopath fails to appreciate the normative force of a demand made on him, he will not view failures to meet the demand as something for which *his* taking responsibility—and all that is implied by taking responsibility, e.g., seeking forgiveness—makes sense. Suppose I don't feel guilt or remorse for having performed some action. There are a couple ways we might explain this. One explanation is that my action did not violate any demands on behavior and thus, while I am capable of guilt and remorse, guilt and remorse would not be appropriate reactions to my behavior. Another explanation is that my action violated a demand on behavior, but while guilt and remorse are thus appropriate, I am incapable of those feelings. Imagine a situation in which a psychopath fails to meet a demand made of him. A psychopath might entertain the idea that he was responsible for not meeting the demand, but he might also know that if he is responsible, then he should feel guilt and remorse. Given that he does not feel these, he can conclude either that he is not responsible after all or that he is highly abnormal. Self-serving thought and behavior is not exclusive to psychopaths, so we would not be baffled if the psychopath concluded the former.

So are addicts and psychopaths not criminally responsible for their behavior? While extreme forms of addiction and psychopathy may render a person not criminally responsible, the implication of *Atkins* is that we need to acknowledge degrees of responsibility. If the features typical of mentally retarded persons render such persons not fully responsible for their actions, and if many of those features characterize addicts and psychopaths, then by parity of reasoning, we should conclude that addicts and psychopaths may not be fully responsible for their actions either. This doesn't mean we need a legal distinction for every difference in degree of responsibility, and for a variety of practical reasons, a proliferation of distinctions would be unwelcomed. However, sole reliance on M'Naghten type tests for insanity tempts us to view those who know the difference between right and wrong as fully responsible for their actions. If those who are found legally insane are not criminally responsible, then criminal responsibility turns out to be an all or nothing affair. This is untenable, and surely our legal system can tolerate a more nuanced conception of responsibility. Put in terms of rationality, the upshot of our present discussion is that we need a more expansive conception of rationality if we want to preserve the connection between rationality and responsibility. If we use only narrow, M'Naghten type tests for rationality, then we will end up treating as rational—and thus responsible—people whose responsibility can be clearly called into question. In turn, this may mistakenly lead us to look for something to plug the gap between rationality and responsibility, e.g., libertarian free will.

With this in mind, recall what Greene and Cohen say about how the law views Mr. Puppet: "The law asks whether or not he was rational at the time of his misdeeds, and as far as we know he was. For all we know, he is psychologically indistinguishable from the prototypical guilty criminal, and therefore fully responsible in the eyes of the law."[35] What sense of "rational" are we assuming here? What is the psychology of the prototypical guilty criminal? In 2006, the Bureau of Justice Statistics found that in state prisons, 73 percent of women and 55 of men had at least one mental health problem;

in federal prisons, 61 percent of women and 44 percent of men had such problems; in local jails, 75 percent of women and 63 percent of men did.[36] If Mr. Puppet is no different from the average guilty criminal, then we have good reason to suspect that Mr. Puppet suffers from a mental illness. Because many mental illnesses do not prevent a person from appreciating the criminality of conduct, the law will treat as rational many people suffering from mental illness. Of course, we do not know enough about Mr. Puppet to assess whether he does suffer from a mental illness, or supposing he does, how that illness might bear on the issue of responsibility. But we are now in a position to see that doubts about his responsibility can be generated by doubts about the conception of rationality the law uses to assess responsibility, and I want to suggest that getting the story about responsibility right turns more on the question of rationality than it does determinism. To illustrate what I have in mind, consider Robert Kane's own hypothetical example of a man on trial, an example Kane uses to raise the question of what ultimate responsibility requires:

> Suppose a young man is on trial for an assault and robbery in which his victim was beaten to death. Let us say we attend his trial on a daily basis. At first, our thoughts of the young man are filled with anger and resentment. But as we listen daily to how he came to have such a mean character and perverse motives—a sordid story of parental neglect, child abuse, sexual abuse, bad role models—some of our resentment against the young man is shifted over to the parents and others who abused and influenced him. We begin to feel angry with them as well as him. Yet we aren't quite ready to shift all of the blame away from the young man himself. We wonder whether some residual responsibility may not belong to him. Our questions become: "To what extent is *he* responsible for becoming the sort of person he now is? Was it *all* a question of bad parenting, societal neglect, social conditioning, and the like, or did he have any role to play in it?" These are crucial questions about free will, and about what may be called *ultimate responsibility*.[37]

For Kane, wondering whether some residual responsibility belongs to the young man is tantamount to wondering whether there is something about him that both escapes the causal nexus and at least partially accounts for the person he has become. But if *this* is what we are wondering about, then the part about abuse and neglect is completely irrelevant. Kane is not concerned about the particulars of the causes at work but their extent, so the wonder will remain regardless of the details of a person's upbringing. It never occurs to Kane that our wondering about responsibility might turn on the perversions wrought by abuse and neglect. When we take seriously that these perversions matter to responsibility, our question becomes "To what extent have these perversions compromised the cognitive and affective capacities necessary for taking responsibility?" This still allows us to pose the issue of "whether some residual responsibility may not belong to him," but the issue becomes not one of factoring contributions to character (e.g., how much is due to genes? How much is due to environment? How much is due to *him*?) but rather of determining the range of the demands we can legitimately make of him.

THE AIMS OF INCARCERATION[38]

Incarceration serves a number of purposes, both punitive and nonpunitive. When incarceration is punitive, the point is to inflict punishment on the person incarcerated. There are a variety of theories that purport to provide a justification for punishment, including retributivist and consequentialist theories. According to Jeffrie Murphy, a retributivist "is a person who believes that the primary justification for punishing a criminal is that the criminal deserves it."[39] We've met with this idea already: a person who takes responsibility for the failure to meet the goodwill demand accepts that she is worthy of punishment, and that such failures merit punishment is just one of the "norms for responsibility."[40] Applied to the law, for a criminal to deserve punishment it seems enough that the person who committed the crime is A-responsible. By contrast, consequentialist theories typically justify punishment by appeal to the socially desirable outcomes that result from punishment, including deterrence. As we saw in Chapter 3, however, the matter is not as straightforward as it may seem. The basic idea behind deterrence is that punishing criminals for their crimes will reduce the likelihood that other people will commit crimes, but there are at least two ways to think about the underlying "logic" of deterrence, and these two ways roughly correspond to the difference between the objective and participant perspective. The first way might go something like this: when I learn about the policy of punishing those who commit crimes, I come to believe that I will be punished if I commit a crime. Given that I have a strong desire to avoid being punished, this belief gives me an incentive to not commit crimes. If this is the logic, note that it seemingly needn't assume that anyone ever deserves punishment; all it requires is that people come to believe that crime leads to punishment. Of course, deterrence can work only if this belief can have some effect on my decision-making, and in this way it seems to assume that I am responsive in ways I've said are characteristic of A-responsible people. Optimists like Greene and Eagleman will grant that deterrence presupposes this kind of responsiveness, but they will deny that this kind of responsiveness implies the kind of moral desert required for retribution. I've argued that they are wrong about this, but we can put the matter aside for the moment and grant that this way of thinking about the logic of deterrence does not assume desert. The second way of thinking about the logic of deterrence goes something like this: when I learn that a person has been punished for a crime, I am reminded of my status as a responsible agent, i.e., I am reminded that certain demands are made of me, and I deserve to be punished when I fail to meet those demands. I don't want to be punished, of course, but my desire to avoid punishment also reflects the fact that I accept certain demands as legitimate. Deterrence results from the fact that the punishment of others reminds me of those demands, thereby making it more likely that those demands will serve as effective reasons. The difference between the two ways of thinking about the logic of deterrence reflects a difference in the way the connection between crime and punishment is conceived. On the first way, the connection between crime and punishment is akin to an empirically discovered regularity: I find out that I live in a world in which punishment reliably follows crime, and because I factor this discovery into my decision-making, I avoid committing crimes. On the second way, the connection between crime and punishment is implied by our conception of the criminal as a responsible agent: The criminal is responsible for her failure and therefore merits punishment.

It is an empirical matter whether punishment is an effective deterrent, of course, and the available evidence suggests that it may not be.[41] However, principled explanations of this ineffectiveness can be given, regardless of whether we adopt the objective or participant perspective. If we adopt the objective viewpoint, we can say that the failure of punishment to be effective simply shows that we were operating with a false empirical theory about human psychology. If we adopt the participant perspective, we can say that it is not the fear of punishment per se that is supposed to deter, but rather the way punishment reminds us that there are legitimate demands on our behavior. If punishment is not in fact an effective means for deterring crime, this might be because punishment is not an effective way of reminding people of these demands. This may seem ad hoc—an alternative explanation is that reminding people of their demands is simply irrelevant, in which case devising better methods of reminding people won't make a difference. But there is evidence that suggests reminding people does make a difference. For example, we might think that the threat of failure, suspension, or expulsion will deter students from cheating. As it turns out, there is an effective means to deter cheating that does not involve such threats. In an experiment conducted by Dan Ariely, it was found that when students have an opportunity to cheat on a test, students who are reminded of moral demands just prior to taking the test cheat far less than students who are not so reminded.[42] From the objective viewpoint, discovering that reminding students of moral demands deters cheating is no different in principle from discovering, say, that taking a certain drug deters cheating. As an empirical matter, this may be right. But again, there is a kind of understanding provided by the participant perspective that the objective viewpoint lacks, one that is related to reasons for action. Cheating avoidance is made intelligible by the reminder students get because they conceive themselves as moral agents who are responsible for their conduct. Taking a wonder pill may also reduce cheating, but the connection will appear seemingly arbitrary; do I see myself as having a *reason* not to cheat when I take the pill? Suppose the pill works by causing you to see yourself as a moral agent who has a reason not to cheat. In that case, we haven't really abandoned the participant perspective on deterrence; rather, we've just found an effective means to create A-responsible people.

The upshot of this discussion is that there is a way of thinking about deterrence that is internally connected to the notion of desert. As such, we shouldn't simply assume that deterrence is an aim of incarceration that is logically distinct from retribution. The thought that it is logically distinct might be explained by the fact that there is a way of thinking about the logic of deterrence that doesn't presuppose desert. While true, there is also a way that does. This difference in the underlying logic of deterrence also suggests that punishment grounded in desert is the only nonaccidentally punitive form of incarceration, for if we reject desert and accept that the connection between punishment and deterrence is contingent, it may turn out that only nonpunitive forms of incarceration are justified.

Nonpunitive aims of incarceration include incapacitation and rehabilitation/treatment. Perhaps the most familiar examples of incapacitation are the so-called 'Three-Strikes' laws that have been enacted by a number of states.[43] These laws mandate that if a person convicted of a felony has prior felony convictions on her record, then this person must receive a longer sentence than she would have if she had not had prior felony convictions. The rationale for the longer sentence thus has nothing to do with the nature of the crime itself or the intentions of the perpetrator at the time the crime was committed. Rather, repeat offenses provide evidence that the person poses a credible threat to public safety, and the point of the longer sentence is to prevent the person from committing further crimes. In this way, incapacitation is not importantly different from other forms of quarantine, e.g., confining a person who has a highly contagious disease in order to prevent others from contracting the disease. If the aim of quarantining a person who has a highly contagious disease is not to punish the person, then parity of reasoning requires that the aim of incapacitation per se be nonpunitive. This isn't to say that incapacitation rules out other grounds for incarceration, of course, and it may be that many of those who are sentenced under Three Strikes laws are properly thought to be responsible for their crimes and thus deserve punishment. But it is important to keep these aims distinct, for if a person is incarcerated on grounds of incapacitation only, then any deprivations beyond those required to protect others would seem not to be justified. Reflection on the person who has a highly contagious disease makes this point clear: protecting others requires that the person be confined, and though confinement results in a loss of liberty (a deprivation), we cannot justify additional deprivations if such deprivations are not required to protect others, e.g., I may need to lock you up in a hospital room, but this does not justify taking away the television set. As we will see below, this has important consequences for how we should think about sentencing.

The other nonpunitive aim of incarceration is rehabilitation and/or treatment. The rationale for incapacitation does not require that we see the person as a legitimate target for rehabilitation or treatment, e.g., we might be convinced that the person is simply not amenable to treatment or cure. Similarly, the rationale for rehabilitation or treatment does not imply that the person we are trying to rehabilitate or treat can be incarcerated on grounds of incapacitation. Of course, given that the person whom we are trying to rehabilitate or treat has committed a crime, we might think that the person poses a threat to public safety. This makes a kind of sense, because to see someone as in need of rehabilitation or treatment is to see that person as not fully responsible, and this implies that the capacities necessary for A-responsibility are compromised to a certain extent. The point here is just that having compromised capacities does not automatically entail greater criminality, a point that the Supreme Court made in *Atkins*. And finally, because less than fully responsible does not mean not at all responsible, rehabilitation or treatment does not rule out there being other grounds for incarceration.

SENTENCING

My discussion of sentencing assumes contexts in which a person is A-responsible for her crime and thus deserves punishment. But the relationship between desert and sentencing isn't straightforward, for at least two reasons. First, saying that a person deserves punishment does not commit us to any specific forms of punishment, including imprisonment. But while there are no strict guidelines governing punishment, we perfectly understand and accept that there are circumstances in which punishment is merited or earned. This is the core concept we need to preserve. Moreover, it may be true that criminals deserve punishment and yet the way we actually punish people is deeply flawed. However, from the fact that our practices of punishment may be deeply flawed it does not follow that the concept of punishment is deeply flawed. Second, because desert need not be the sole ground for incarceration, we cannot assume that the appropriateness of a sentence is a direct function of desert.

Consider life sentences. Do those who are sentenced to life deserve to spend the rest of their lives in prison? Some life sentences come with the possibility of parole. If a person sentenced to life deserves to spend the rest of his life in prison, then parole would look like an injustice. A possible reply is that the sentence is a function of desert only, but desert itself is in part a function of how one actually responds to the violation. In satisfying the conditions for parole, a person no longer deserves to be in prison. The conditions for parole include good behavior and remorse, and often require that the prisoner demonstrates a sustained commitment to bettering himself. If a person does not do this, then he deserves to stay in prison. Here, deserving life is the default assumption, but this can be revised in light of subsequent behavior. Let's call this the *strict-but-fluid desert* position. Another possible reply is that the sentence is a function of desert and other considerations, e.g., incapacitation. Perhaps the desert includes only the time before the possibility of parole, so that regardless of behavior, the person no longer deserves to be in prison after that time is up. The person is not automatically released, however, because the person might be judged too great a threat to public safety. According to this position, the point of satisfying the conditions for parole is to demonstrate that the person is no more dangerous than the "average" person. Let's call this the *desert-plus* position.

The difficulty with strict-but-fluid desert is that some life sentences do not come with the possibility of parole, but a person so sentenced can act the very same way that a person who is paroled acts. Put another way, we cannot automatically assume that the person who is sentenced to life without parole will not act like the successfully paroled person. Not only that, a person sentenced to life without parole may reform themselves in ways that those sentenced to life with the possibility of parole actually do not. Between the two, the former would seem to be more deserving of release than the latter, and yet that difference would be cancelled out; in the former case, parole is not possible, and in the latter case, the conditions for parole are not met. If desert is partly a function of subsequent behavior, then strict-but-fluid desert seems to entail that life without the possibility parole is not fair.

Perhaps the defender of strict-but-fluid desert will argue that some crimes are such that nothing one does can affect whether the person deserves to be in prison. For these kinds of crimes, *strict-and-solid* desert supersedes strict-but-fluid desert. The problem with this is that it is evidently ad hoc. In the case of life with the possibility of parole, we are granting that desert isn't just about the crime committed, but also subsequent behavior. And presumably this point applies to lesser crimes, crimes that do not result in life sentences but also include the possibility of parole (e.g., theft, assault, etc.). Now we are saying that sometimes desert is a function of the crime only. But why? Maybe it will be argued that crimes that result in life without the possibility of parole are the worst crimes possible, and that's what makes them different. But here it seems like we have a straightforward way of explaining the sentence without invoking the notion that in some cases the nature of the crime itself fixes the desert. Invoking desert-plus, we can say that some crimes are such that no amount of behavioral evidence will convince us that releasing the person is worth the risk. This is a policy question, not a metaphysical or epistemological one, but the net effect of this policy is to make the question of desert somewhat irrelevant. Because the person is A-responsible, the person deserves some prison time. But we don't have to worry about how much time the person deserves (regardless of whether desert is a function of subsequent behavior) for even if the person no longer deserves to be in prison, there are other reasons to keep the person incarcerated, e.g., incapacitation. The defender of desert-plus will thus see the nature of the crime as relevant to the standard of risk we apply when considering whether to release a prisoner. Crimes which result in life without the possibility of parole differ from the crimes which result in life with the possibility of parole in that the standards of risk we apply in the former case, but not the latter, are such that no one can feasibly meet them.

To illustrate the basic idea behind desert-plus, consider Major League Baseball's long-standing policy of permanently banning players and coaches who bet on professional baseball games. In 1919, several Chicago White Sox players took part in a gambling conspiracy to throw that year's World Series. This violation of public trust was so serious that many believed it threatened the institution of professional baseball itself, and newly appointed commissioner Kenesaw Mountain Landis responded by permanently banning from baseball those who participated in the conspiracy. Gambling is not the only activity thought to compromise the integrity of professional sports, however. In the first decade of this century baseball confronted a series of scandals involving the use of illegal performance enhancing substances, including steroids. But while gambling and the use of steroids each violates Major League Baseball policy, the response to such violations has tended to be quite different. Precedent tells us that gambling results in a lifetime ban, but most drug policy violations result in suspension or fine. Why do these violations meet with different results? Major League Baseball is a business enterprise whose existence depends on the financial support of fans and advertisers. While there may be countless reasons why fans find interest in the game, there can be no doubt that this interest is predicated on the belief that each team is trying to win. Gambling undermines this belief and thus undermines the fan interest needed to sustain the game. In this way, gambling poses an existential threat to the game itself, and so we can make sense of why someone who violates the gambling policy would receive a permanent ban; even if the probability of a second offense were low, and even if the low probability would be explained by the fact that the violator took responsibility for the violation, the risk of allowing the violator to continue participating is simply too great. Correctly or incorrectly, the use of steroids is not thought to pose this kind of threat, and so while such use may warrant punishment, it does not disqualify membership.

I've said that in the ideal case in which a crime is committed, the person who committed the crime will take responsibility for the crime and seek forgiveness. As we saw in Chapter 3, one of the ways we punish a person who violates the goodwill demand is by withdrawing our goodwill toward that person, and when we are prepared to forgive, we are prepared to reinstate this goodwill. In the context of law, our forgiveness is signaled by our willingness to release a person from prison. But does it make sense for a criminal to take responsibility and seek forgiveness in cases where the criminal cannot be forgiven, e.g., in cases where the criminal is sentenced to life without the possibility of parole? Consider a first-degree murderer who is sentenced to life without the possibility of parole. The person might take responsibility for his crime, and as a result of taking responsibility, the person might bring about changes in himself that would keep him from committing further crimes if he were to be released. First-degree murderers can be A-responsible, and so in that way it can make sense to take responsibility. However, I've also argued that A-responsibility is a mechanism by which D-responsible agents are produced, and our interest in the existence of such agents is predicated on the assumption that the behavior of other people can significantly impact our lives. Thus a person facing life in prison without the possibility of parole might reasonably conclude that it doesn't much matter what his future behavior is like. Practically speaking, taking responsibility will look pointless.

Because desert-plus says that certain crimes justify life sentences without the possibility of parole, it looks like desert-plus also says that certain crimes are unforgiveable. But this is misleading. Desert-plus is not committed to saying that certain crimes are unforgiveable; it is committed to saying that some crimes won't be forgiven. The difference is important. That certain crimes won't be forgiven reflects a pragmatic policy decision about which risks we are, or are not, willing to take. All that matters to justifying this policy decision is that we—including the criminal—understand why certain risks are not worth taking. Because desert-plus appeals to risk to account for life sentences, we can even imagine situations where it might make sense to forgive, e.g., we might forgive a prisoner who is on the cusp of death, on the grounds that he can no longer pose a threat. The underlying rationale here is thus significantly different from the one at work in strict-and-solid desert. According to that view, some crimes are such that it is *impossible* to justify forgiving those who commit them. Imagine a Raskolnikov so guilt-stricken that a moral saint emerges, like a phoenix, from the ashes of his former self. Unlike desert-plus, considerations of risk are not relevant to lifetime imprisonment, and the defender of this position can allow that we might know with absolute certainty that our moral saint would never commit another crime. It simply doesn't matter, even if our moral saint is on death's door. If desert-plus makes taking responsibility practically pointless, strict-and-solid desert makes taking responsibility metaphysically absurd.

Perhaps it will be thought that the difference between being practically pointless and being metaphysically absurd is a difference that doesn't make a difference. However, desert-plus faces a complication that strict-and-solid desert does not, but a complication that also suggests a way to restore the point of taking responsibility. As a stand-alone justification for incarceration, incapacitation does not justify depriving a person of something beyond what is required to protect others. Suppose a person commits assault in a state that enforces a Three Strikes law, and the person has two prior felony convictions on her record. It might be that the person faces a mandatory life sentence, even if the person does not deserve a life sentence for committing assault. For the sake of argument, assume that the person deserves 25 years in prison for having committed the assault. If prison deprives the person of something beyond what is necessary to protect others, then it would seem that we cannot justify *imprisoning* this person beyond the first 25 years. To think otherwise is to collapse the distinction between desert-plus and strict-and-solid desert. But now consider what might happen if we took the distinction between desert and incapacitation seriously. Perhaps the person would leave prison and be detained under some form of house arrest. Though still incarcerated, release from prison would signal that the person has been forgiven, and in forgiving her, we acknowledge that she can be a good standing member of the community—with the proviso, of course, that the conditions for protecting others is met. As a corollary, despite the fact that house arrest may be a permanent condition for this person, it would make sense for her to take responsibility and seek forgiveness, and while the condition of house arrest that results from applying the higher standard of risk may affect the kind and quality of relationships that the person can enter into, it is presumably a vast improvement over prison.

Admittedly this may all sound a bit odd, but the oddity results from the fact that we are operating with a different standard of risk. A different, albeit related, example might remove some of the oddness. Suppose a drug addict commits a crime, and let's assume that the addiction was relevant to the crime, e.g., the person had no money and robbed a liquor store in order to buy drugs. The person goes into recovery while in prison and is scheduled to be soon released. Let's stipulate that it is highly likely that the person will relapse if we let him out and it is highly likely that he will commit future crimes if he relapses (but not otherwise). In light of these considerations, we might think we have a good reason to keep the person in prison, e.g., incapacitation. But what if there were steps we could take that would practically guarantee that the person would not relapse if released? All else being equal, if the resulting deprivations were fewer than the deprivations incurred in prison, then assuming we could not appeal to desert to justify further prison time, it would seem that we should release the person and take those steps.

I've argued that determinism does not threaten the core commitments of the participant perspective, including responsibility and desert. Following Strawson, I've also argued that the concepts of responsibility and desert are inextricably linked to a host of attitudes and practices, including the practices of taking responsibility and seeking forgiveness. In the context of law, I put this point by characterizing the ideal case in which a person commits a crime. In such a case, the person will acknowledge that she failed to meet a legitimate demand and thus ought to seek forgiveness. That a person ought to seek forgiveness does not imply that we will forgive her, of course, but it does imply that forgiveness is possible. Because strict-and-solid desert is committed to saying that certain crimes *cannot* be forgiven, I think we should reject such a view. However, I don't presume this to be the final word on the matter, and it is enough for my current purposes to draw attention to questions—urgent questions—we should be asking. To underscore the urgency of these questions, let me close with a final example.

At the time of writing this, John MacKenzie has spent 40 years in prison for the 1975 killing of a police officer named Matthew Giglio. MacKenzie was sentenced to life in prison, with the possibility of parole after 25 years. According to an editorial in the *New York Times*

> No one disputes that Mr. MacKenzie has been a model prisoner with a perfect record … . He has counseled other inmates nearing their own release dates, and, with the help of a $10,000 grant, he established a program to give victims an opportunity to speak directly to inmates about the impact of their crimes. At his last parole hearing, he told the board, "I did that in memory of Matthew, to show his family that this is the best I can do to make up for it." He said that while he was high on drugs during his crime and did not remember shooting Mr. Giglio, the murder was "a hundred percent my fault."[44]

Though he has been eligible for parole for the last 16 years, his parole has been consistently denied. According to the Parole Board, "Mr. MacKenzie's crime showed 'a serious disregard for the law' and for 'the sanctity of human life,'"[45] and "his release would 'undermine respect for the law.'"[46] However, "The decision did not say Mr. MacKenzie posed any risk to society,"[47] a response that is "typical for the Parole Board, which routinely ignores state laws and regulations that require it to consider an inmate's rehabilitation, and to use a sophisticated risk-assessment tool to help determine whether it is safe to return that person to the community."[48] The State Supreme Court has held the Parole Board in contempt for perceived failures such as this, and this might "help force the board to use parole as intended: not to minimize the seriousness of a crime, but to acknowledge that the person who committed it can change and deserves a chance to rejoin society."[49] The position of the Parole Board seems to reflect strict-and-solid desert; yes, Mr. MacKenzie has taken responsibility for his crime; yes, Mr. MacKenzie has reformed himself; yes, Mr. MacKenzie poses no threat to society. *But that doesn't matter.* However, according to the New York State Supreme Court and *The New York Times*, it *does* matter. Our challenge is to figure out who is right. I will revisit this issue—and many of the other issues raised in the last three sections—in Chapter 9, but it is now time to confront challenges for law posed by the use of neuroscientific evidence.

NOTES

1. Robert Cummins and Martin Roth, "Traits Have Not Evolved to Function the Way They Do Because of a Past Advantage," in *Contemporary Debates in Philosophy of Biology*, ed. Francisco Ayala and Robert Arp (Malden, MA: Blackwell, 2010), 79.
2. Stephen Finlay, "Oughts and Ends," *Philosophical Studies* 143 (3) (2009): 315–340.
3. Cummins and Roth, "Traits."
4. Jaegwon Kim, *Supervenience and Mind* (New York: Cambridge University Press, 1993).
5. Carl Craver, "Functions and Mechanisms: A Perspectivalist Account," in *Functions: Selection and Mechanisms*, ed. Philippe Huneman (Dordrecht: Springer, 2013), 133–158.
6. Cummins and Roth, "Traits."
7. Or, even if we think it can be achieved within the participant perspective, we might also think that it would be too time consuming or less efficient than simply putting Douglas on Ritalin.
8. For excellent discussions of these issues, see Philip Kitcher, *Science, Truth, and Democracy* (New York: Oxford University Press, 2001) and Kwame Appiah, *Cosmopolitanism* (New York: W.W. Norton & Co., 2006).
9. I suspect that these kinds of considerations figure significantly in the thinking of social constructionists. For a nice treatment of this issue, see Ian Hacking, *The Social Construction of What?* (Cambridge, MA: Harvard University Press, 2000).
10. Arthur Samuel Jones, *Thoreau's Incarceration, as Told by His Jailer* (Berkeley Heights, NJ: Oriole Press 1962), 15.
11. Daniel Dennett, "My Brain Made Me Do It." Max Weber Lecture Series, European University Institute. Florence, Italy, December 15, 2010, accessed August 30, 2016, http://cadmus.eui.eu/bitstream/handle/1814/16895/MWP_LS_2011_01.pdf?sequence=1, 11. For a thorough development and defense of an idea along these lines, see Manuel Vargas' *Building Better Beings* (New York: Oxford University Press, 2013).

12. Trivers, "Reciprocal Altruism"; Richard Dawkins, *The Selfish Gene* (Oxford: Oxford University Press, 1989); Robert Axelrod and William Hamilton, "The Evolution of Cooperation," *Science* 211 (1981): 1390–1396.

13. The fly in this case is the Prisoner's Dilemma. An excellent discussion of this dilemma is provided by William Poundstone, *Prisoner's Dilemma: John von Neumann, Game Theory, and the Puzzle of the Bomb* (New York: Doubleday Anchor, 1992).

14. I assume self-regarding actions are social in the relevant respect on the grounds that, like all actions, they involve both and an agent and a patient.

15. Recall that this is Strawson's characterization of what is involved in seeking forgiveness.

16. This matter is more complicated in cases of life sentences that do not come with the possibility of parole. I return to this issue later in the chapter.

17. Richard Moran, *Knowing Right from Wrong: The Insanity Defense of Daniel McNaughtan* (New York: Free Press, 1981), 173. Moran is quoting Chief Justice Tindal's majority opinion.

18. "Illinois Compiled Statutes," Illinois General Assembly, accessed September 5, 2016, http://www.ilga.gov/legislation/ilcs/fulltext.asp?DocName=072000050K6-2.

19. Each of the two readings I provide here assumes that repeat criminality does not entail the lack of substantial capacity. However, there is another way to read (b) in which it does. Suppose repeat criminality does entail lack of substantial capacity. If so, then what (b) says is that lacking the capacity is not enough for insanity, even if it is the result of disease or defect. To be criminally insane, one must lack other capacities as well, e.g., the capacity to solve reasoning problems, or the capacity to correctly identify different types of objects, or whatever. As a strictly logical matter, this condition would seem completely unmotivated. Suppose we wanted to define what it is to be blind. Parroting (a), we might say that a person is blind if, as a result of neurological disease or defect, the person is unable to see. If we read (b) as I am considering it now, then we would be saying that diseases or defects that manifest only in the inability to see won't count as diseases or defects that would qualify a person as blind. That is patently absurd. As an empirical matter, however, it is probably true that a person who lacks the capacity to appreciate the criminality of his behavior also lacks other capacities as well. The requirement that the disease or defect manifest in other ways would thus likely not have undesirable practical consequences.

20. *Durham v. United States*, 214 F.2d 862, 874–75 (D.C. Cir. 1954).

21. *United States v. Brawner*, 471 F.2d 969 (D.C. Cir. 1972).

22. *Model Penal Code 4.01*, American Law Institute (1962).

23. 21 Am.Jur.2d *Criminal Law* 38 (1998).

24. 40 Am.Jur.2d *Homicide* 109, n. 2 (1999).

25. 21 Am.Jur.2d *Criminal Law* 38 (1998).

26. *People v. White*, 117 Cal. App. 3d 270 (1981).

27. *United States v. Hinckley*, 672 F.2d 115 (D.C. Cir. 1982).

28. *Atkins v. Virginia*, 536 U.S. 304 (2002).

29. For states that use a version of the M'Naghten Rule, the practical implications of *Atkins* will likely show up only during the sentencing phase, e.g., as a factor in a mitigation argument. I'll return to this point in Chapter 9.

30. "Definition of Addiction," American Society of Addiction Medicine, accessed September 30, 2016, http://www.asam.org/quality-practice/definition-of-addiction.

31. Scott Lilienfeld and Hal Arkowitz, "What 'Psychopath' Means," *Scientific American*, December 1, 2007, accessed October 3, 2016, https://www.scientificamerican.com/article/what-psychopath-means/.

32. For a nice exploration of what the details of this integration might look like, and what these details might imply for the law, see Katrina Sifferd, Tyler Fagan, and William Hirstein's "Legal Insanity and Executive Function," in *The Insanity Defense: Multidisciplinary Views on Its History, Trends, and Controversies*, ed. Mark White (Santa Barbara, CA: Praeger, 2017), 215–242.

33. Strawson, "Freedom and Resentment," 86.

34. Strawson writes, "The making of the demand *is* the proneness to such attitudes" ("Freedom and Resentment," 90). If that is correct, then perhaps it would be more accurate to say that the psychopath is incapable of seeing demands *as* demands. In such case, we might say psychopathy is a condition in which the only available viewpoint toward the world is an objective one.

35. Greene and Cohen, "Neuroscience Changes Nothing and Everything," 1780.

36. Doris James and Lauren Glaze, "Mental Health Problems of Prison and Jail Inmates," *Bureau of Justice Statistics* NCJ 213600 (2006).

37. Kane, "Free Will," 501.

38. It is beyond the scope of this book to do justice to the vast philosophical literature on incarceration and punishment, and while seminal work in the theory of punishment—including the retributivism of Immanuel Kant (*The Philosophy of Law: An Exposition of the Fundamental Principles of Jurisprudence as the Science of Right* (Union, NJ: Lawbook Exchange, 2002)), the consequentialism of Jeremy Bentham (*An Introduction to the Principles of Moral and Education* (London: Methuen, 1982)), and mixed theories along the lines developed by H.L.A. Hart (*Punishment and Responsibility* (New York: Oxford University Press, 1968))—clearly provides the philosophical foundation for much of my discussion, the primary aim in this section (and the next) is to consider well established positions on incarceration and punishment (developed by the aforementioned thinkers, among others) through the lens of responsibility developed in this book. For an excellent overview of the philosophical literature on punishment, see Antony Duff's "Legal Punishment" (*The Stanford Encyclopedia of Philosophy*. Accessed August 4, 2017, https://plato.stanford.edu/entries/legal-punishment/).

39. Jeffrie Murphy, "Legal Moralism and Retribution Revisited," *Criminal Law and Philosophy* 1 (2007): 11.

40. Dennett, "My Brain Made Me Do It," 11.

41. A nice discussion of this issue can be found in Raymond Paternoster, "How Much Do We Really Know about Criminal Deterrence?" *Journal of Criminal Law and Criminology* 100 (3) (2010): 765–824.

42. Dan Ariely, *Predictably Irrational* (New York: Harper Collins, 2009).

43. A detailed discussion of these laws can be found in Franklin Zimring, et. al., *Punishment and Democracy: Three Strikes and You're Out in California* (New York: Oxford University Press, 2001).

44. "A Challenge to New York's Broken Parole Board," *New York Times*, June 13, 2016, accessed July 22, 2016, http://www.nytimes.com/2016/06/13/opinion/a-challenge-to-new-yorks-broken-parole-board.html?_r=0.
45. *New York Times*, "Challenge."
46. *New York Times*, "Challenge."
47. *New York Times*, "Challenge."
48. *New York Times*, "Challenge."
49. *New York Times*, "Challenge."

Chapter 8

fMRIs Are Watching You

If the machinery of free will and responsibility is implemented in the brain, then observation of the brain may reveal important details about this machinery and the conditions that can disrupt its functioning. Generalizing, if our mental life is rooted in our biology, then observing the brain may tell us something about a person's mental life. Thus while questions about the capacities required for responsibility are of special interest to the law, observation of the brain may also have important implications for detecting juror bias, predicting future criminality, assessing the reliability of witness testimony, and determining the potential for rehabilitation or treatment. As we've seen, however, there are at least three positions that satisfy the conditions of physicalism—Cartesian materialism, Ryleanism, and functionalism—and how we conceive the relevance of neuroscience to questions about mind will depend on which position we adopt. This point applies to the results of neuro-imaging as well, and in this chapter I approach the question of relevance from a broadly functionalist perspective.

I've argued that reconciling the manifest and scientific images of humans requires showing how a purely physical system can be a free, rational, and responsible agent, and in Chapter 6 I provided a three-step strategy for doing this. The first step involves characterizing the target phenomena, e.g., free will, as capacities. The second step involves providing a functional analysis of the relevant capacities, i.e., to show that a system with a certain kind of design is bound to have the capacity in question. The third step involves showing how the design is implemented in a purely physical system. From the point of view of functional analysis, the claim that you are your brain is thus the claim that the design in question is implemented in the brain. There are two important implications of this for the use of neuro-imaging, as I will now discuss.

BRAINS AND BEHAVIOR

Suppose we are interested in figuring out the extent to which adolescents are rational. If you are your brain, then we might think that we can answer this question by observing adolescent brains. But as Greene and Cohen rightly point out,

> This gets the order of evidence backwards. If what the law ultimately cares about is whether adolescents can behave rationally, then it is evidence concerning adolescent behaviour that is *directly* relevant. Studying the adolescent brain is a highly *indirect* way of figuring out whether adolescents in general are rational. Indeed, the only way we neuroscientists can tell if a brain structure is important for rational judgement is to see if its activity or damage is correlated with (ir)rational *behaviour*.[1]

In conceding the primacy of behavior, we should acknowledge this kernel of truth in Ryleanism: terms like "rational," "impulsive," and "insane" pick out behavioral capacities, not brain states, so if we want to know whether a person is rational, impulsive, or insane, observations of a person's behavior are obviously relevant. The point isn't that observing the brain is irrelevant to this project, and later on I will discuss how results from neuro-imaging can contribute something important here. Rather, the point is that observing behavior is often our best source of evidence regarding behavioral capacities. For example, suppose I want to know whether a certain car is fast, and I stipulate that a car is fast if it can go from 0 to 60 miles per hour in less than six seconds. Here's a way to do it: turn on the car, start the clock, and hit the gas. Note, too, that judgments of rationality long predate neuro-imaging technologies, and it is plausible to think that at least some of those judgments have been well founded. We can even go as far as saying that the relationship between behavior and rationality is "conceptual." Imagine devising the most sophisticated behavioral test for rationality possible, and suppose that a person convincingly passes the test. Does it make sense to ask whether the person is *really* rational? Put another way, is it genuinely possible that a person be irrational if, in principle, *no* behavioral test would reveal it? I don't think one has to be the village behaviorist to be suspicious of such a possibility.

Perhaps it will be objected that, in *practice*, behavioral evidence typically underdetermines the presence of a capacity, and the use of neuro-imaging would reduce the uncertainty of the inferences we draw from behavior alone. Consider the capacity to multiply. If a person provides the correct solution to a multiplication problem, it does not follow that the person has the capacity to multiply; the person may have made a lucky guess, or the person may have learned to solve this problem by rote. Conversely, if a person fails to provide the correct solution to a multiplication problem, it does not follow that the person lacks the capacity to multiply; the person may have misread the problem, or the person may have made a slip of the pen. As Ryle notes, though capacities are dispositions, most capacities of interest are "not single-track dispositions, but dispositions the exercises of which are indefinitely-heterogeneous."[2] This point is generally appreciated, and in controlled experiments, e.g., the experiments conducted on E.V.R., determining the presence, absence, or extent of a person's capacities often requires observing behavior over time, in a variety of settings and conditions, and on a number of different tasks. This is also reflected in the law, e.g., as we discussed in the last chapter, parole boards use fairly sophisticated risk-assessment tools in the process of deciding whether to grant parole. But if behavioral capacities supervene on brain and body, then it is plausible to think that observations of a person's brain can reveal something about the person's capacities which observations of behavior might not.

Indeed, it is plausible; in fact, we can defend two stronger claims. First, though rationality (or impulsivity, or insanity) can be defined in terms of behavioral capacities, and though behavioral evidence can justify calling a person rational, justification for calling a person rational (or impulsive, or insane) does not *require* behavioral evidence. For example, given enough information about a car's engine, transmission, weight, shape, and so forth, I can reach a reasonable conclusion about performance capacities even if the car never leaves the garage. Similarly, detailed knowledge of the brain can reveal how a person would behave in various settings, and thus I might come to know whether a person is rational even if I never observe her behavior. Second, the possibility of coming to know about capacities in this way seems to turn on the fact that the information which grounds my inference about capacities is information which *explains* the capacities. Consider again the capacity to multiply. Knowing that a particular machine implements a partial products algorithm can justify the conclusion that the machine has the capacity to multiply, but knowing that the system implements a partial products algorithm is also what allows me to explain the capacity to multiply. If the knowledge that would justify attributing rationality is also knowledge that would explain rationality, then results of neuro-imaging look poised to provide explanatory benefits in a way that mere observation of behavior does not.

But would these alleged benefits matter to law? In the last chapter we met with the idea that successful insanity defenses typically require showing that the substantial lack of capacity to appreciate criminality of conduct is the result of disease or defect, and if disease or defect is meant to explain the lack of capacity, then it seems that the explanatory benefits would matter to law.[3] However, we also noted two difficulties with the appeal to disease or defect. The first was that our grounds for attributing disease or defect rest on considerations of behavior, i.e., it is the fact that a person lacks the capacity in question that would justify saying that the person has a disease or defect. If there is no way of independently identifying disease or defect, then disease or defect wouldn't explain the lack of capacity; rather, talk of disease or defect would amount to no more than a redescription of the fact that the person lacked the capacity in question. The second difficulty was that it is hard to motivate the claim that explanatory factors matter; what matters is that the person lacks the capacity in question, not why the person lacks the capacity. Or rather, the explanatory factors matter only insofar as they play an important evidentiary role, i.e., those factors are what *convince us* that the person lacks the capacity. These points apply to the issue of future behavior as well, e.g., recidivism. In deciding whether to release someone from prison, a parole board wants to know about the prisoner's behavioral tendencies, not why the prisoner has those tendencies. Explanatory factors might be relevant here, but again their relevance is largely evidential. Imagine a person who is known to be an infallible source of information about behavioral tendencies. The person testifies at a parole board hearing, declaring simply that the prisoner who is up for parole poses no credible threat to others. Despite the fact that the testimony does not explain why the prisoner poses no credible threat, the parole board comes to have a conclusive reason for thinking the prisoner poses no credible threat. *That* is what matters to the parole board's decision.

However, there is at least one area of law where the explanatory benefits clearly seem relevant. If rehabilitating criminals is a legitimate aim of the criminal justice system, then our interest cannot simply be to know whether a person lacks a capacity. Such knowledge is of course relevant to whether we *view* someone as a candidate for rehabilitation, but this is not sufficient. One reason is obvious enough; knowing that someone lacks a capacity tells us virtually nothing about how the capacity might be restored or enabled. By analogy, while behavioral evidence might convince me that my car won't start (I turn the key and nothing happens), diagnosing and fixing the problem requires looking under the hood. Moreover, whether we view a person as a suitable candidate for rehabilitation depends on whether we believe there are feasible steps we can take to restore or enable the capacity in the first place, and whether we believe there are such steps will depend on what we take the explanation of the lack of capacity to be. For example, suppose a person lacks a certain capacity as a result of suffering from schizophrenia. Because there are a number of drugs (e.g., clozapine and risperidone) that have proven to be quite effective in treating schizophrenia, we might plausibly conclude that such a person can be rehabilitated.[4] By contrast, if a person lacks a certain capacity as a result of suffering significant damage to the prefrontal cortex, then absent existing medical techniques for repairing such damage, we might plausibly conclude that such a person cannot be rehabilitated (or at least not in the near future). Thus in coming to know why a person lacks a capacity, we may come to know whether rehabilitation is possible, and if it is possible, what steps we might take to restore or enable the capacity.

I've argued two claims: (a) in principle, results from neuro-imaging may eliminate the need for behavioral tests altogether; (b) results from neuro-imaging can provide explanatory benefits that behavioral tests might not. But I also want to be clear about what I've *not* argued. First, I have not argued that results from neuro-imaging provide "direct" evidence of rationality (or impulsivity, or insanity, etc.), whereas behavioral tests only provide "indirect" evidence. To be rational is to possess certain behavioral capacities, and evidence for such capacities can come from observing behavior and from observing the brain-plus-body on which these capacities supervene. The idea that neuro-imaging might provide "direct" evidence for rationality rests on the assumption that rationality is a brain state, but even most Cartesian materialists don't believe this.[5] Second, I have not argued that neuroscientific evidence is "hard" and "objective," whereas behavioral evidence is "soft" and "subjective." It may be that a lot of people believe this, including jurors,[6] but while this belief may be relevant to the rhetorical success of arguments presented in courtrooms, nothing I've said supports this belief. Finally, I have not argued that we are anywhere close to replacing behavioral evidence with evidence gathered from neuro-imaging. Such a replacement is possible, but justifying such a replacement requires us to deal successfully with the second important implication of functional analysis for neuro-imaging.

REVEALING IMAGES[7]

Ronald Giere provides a useful way to think about the role of neuro-imaging in generating observations of the brain.[8] Like our five senses, scientific instruments respond to a limited range of features of their environment.[9] Thus while we may speak of neuro-imaging techniques as enabling us to observe the brain, it would be more correct to say that these techniques enable us to observe certain *aspects* of the brain. Which aspects these techniques enable us to observe will depend on the kind of instrument used. For example, images produced by computer assisted tomography (CAT) represent what Giere calls *structural* features of brains, while images produced by positron emission tomography (PET) represent what Giere calls *functional* features of brains.[10] Images produced by functional magnetic resonance imaging (fMRI) represent both structural and functional features.[11] More specifically, fMRI can reveal locations where oxygenated or deoxygenated blood dominates. The sense in which this reveals functional features is that, on the assumption that the use of oxygen correlates with brain activity, fMRI can reveal regions of higher and lower brain activity during the performance of some task. However, although knowing that a particular area of the brain is highly active during the performance of some task might provide a good reason for thinking that this area of the brain contributes something relevant to the task, there are at least three factors that complicate such inferences. First, in nervous systems we find organized structures at many spatial scales, including individual neurons, networks comprising neurons, and systems comprising networks.[12] Assuming that functions can be ascribed to these organized structures, we can ask *at what scale* does an image purport to reveal structural or functional features? An image might reveal structural or functional features of a neuron without revealing structural or functional features of the networks and systems of which the neuron is a part. Conversely, an image might reveal structural or functional features of a system without revealing structural or functional features of the networks and neurons that make up the system. In the context of functional analysis, the functional features that are relevant will depend on the capacity we are trying to explain. For example, suppose the visual system produces representations of the spatial layout of the surrounding environment, and suppose that processing information about the spatial layout of the surrounding environment is part of the explanation of a person's capacity to find her way home. In this case the functional features of the neurons that make up the visual system are irrelevant to the explanation, even if we grant that those features are relevant to the explanation of how the visual system manages to produce such representations. But now imagine that the only information neuro-imaging provides is information about the activity of individual neurons. In that case, though neuro-imaging might reveal that a particular neuron is active when the person heads home—and thus reveal functional features, in some sense—it wouldn't reveal the functional features relevant to the explanation of the target capacity.

A second factor complicating the conclusions we can draw from neuroimaging is that functional discreteness does not entail physical discreteness (and vice versa). We met with this point in Chapter 6, in connection with the discussion of scattered causation. An image produced by fMRI might reveal that physically discrete parts of the brain are simultaneously active during the performance of some task, but while physical discreteness can correspond to functional discreteness, physical discreteness does not entail functional discreteness. Recall Lycan's example of how, relative to the word processing capacities of a conventionally programmed home computer, sentences are stored in functionally discrete but physically scattered ways, e.g., a sentence may be stored in widely disparate and "inscrutably linked" memory addresses. As Lycan notes, the links can be *so* inscrutable that an expert technician might not have a clue where the computer has stored a particular sentence. For present purposes, we can put the point as follows: from the fact that (a) information is stored in memory address A, (b) information is stored in memory address B, and (c) A and B are physically discrete, it does not follow that the information stored in memory address A is *functionally* discrete from the information stored in memory address B (again, relative to the word processing capacities of the computer). If we replace memory addresses with neural networks and replace information with activity, then even if the collective activity of networks A and B is functionally discrete, i.e., it contributes something specified by a functional analysis of a target capacity, it need not follow that the activity of the networks *considered individually* is functionally discrete.

The explanatory targets of functional analyses are capacities, and we are granting that to call someone rational, impulsive, or insane is to say something about the person's capacities, not states. But this does not require us to view all talk of the mental as talk of capacities, e.g., we can treat various of a person's thoughts, feelings, moods, and perceptions as states of the person.[13] If the identity theory is true, i.e., if mental states are brain states, then some images of brain states will be images of mental states.[14] But just as an image of a brain represents some of its features but not others, an image of a brain state represents some of its features but not others. This leads to the third factor complicating the conclusions we can draw from neuro-imaging. Recall the analytic functionalist's strategy for deriving the identity claim 'pain = c-fiber stimulation': the concept of pain is the concept of an inner state that is caused by stabs and kicks and causes screaming and writhing. C-fiber stimulation satisfies the causal profile for the concept pain, i.e., it is the state that is caused by stabs and kicks and causes screaming and writhing. Therefore, pain = c-fiber stimulation. Now suppose that I am presented with an image that depicts c-fiber stimulation. What aspect of c-fiber stimulation must the image depict in order for me to conclude, on the basis of the image alone, that the image is an image of *pain*? Obviously, the fact that c-fiber stimulation satisfies the causal profile for the concept of pain. Arguably, no image produced by current neuro-imaging techniques reveals the causal or function profiles of brain states, and if that is correct, then such images do not represent brain states *as* mental states, i.e., they do not reveal the features of brain states in virtue of which such states qualify as mental.

The claim that an image of a brain state does not reveal causal or functional profiles is *not* a claim about current technology. As technology improves, we can reasonably expect that images will provide depictions that are more detailed and accurate than what is available now. In the context of law, whether an image provides good evidence for a particular conclusion might turn on the level of detail and specificity provided by the image, as well as the overall reliability of the technique used to produce the image. More advanced imaging technologies may thus meet standards of evidence that current technologies do not meet. However, such advances mean little in isolation. For example, suppose imaging technologies allow us to depict the workings of a single neuron. Improved technology might reveal ever greater details about the structure of a neuron and the changes a neuron undergoes when it "fires," but such information by itself won't tell us about the *role* the neuron plays in the larger functional economy of the brain.

If an image produced by current imaging technologies does not reveal the causal profile of a brain state, it does not reveal enough information to ground the conclusion that we are observing a mental state. This point applies, mutatis mutandis, to the functionally characterized states, processes or components that figure in the functional analysis of some capacity. This is not to say that images don't or couldn't reveal such profiles, however. To reveal causal profiles, images have to depict inter-neural dynamics (and the way those dynamics are connected to behavior), and as motion pictures demonstrate, *sequences* of images can reveal dynamics. But if the problem with a single image is that it does not reveal enough information, the problem with a sequence of images is that it may reveal *too much* information. Recall that the explanatory value of functional characterization is that it provides causal relevance filters. Being in a certain brain state will have a multitude of causal consequences, but not all of those consequences will be relevant to the capacity we are trying to explain. Absent causal relevance filtering, we are left with noisy detail with no indication of what is relevant and what is a mere by-product of this or that implementation. But there is no reason to think that imaging techniques will reveal only those causal consequences of brain states that are relevant to the explanation of a target capacity, and in this way, images without functional analyses are blind.

The point of these skeptical sounding remarks is not to dismiss the use of imaging techniques or to claim that "observing brain states" is not "observing mental states." The point rather is that, absent functional characterization, images of the brain tell us practically nothing about the mind. Or rather, they tell us practically nothing in the kind of contexts relevant to law. Let me explain. In Chapter 5 I noted the need to distinguish the explanation of a capacity from the confirmation of a functional analysis, and if our task is to confirm (or disconfirm) a proposed analysis, neuro-imaging can of course play a crucial role. We can think of a functional analysis as generating a model of some target system, e.g., the brain, and the task of confirming a model is the task of identifying elements and functional relations in the model with certain states of the brain and causal relations among those states. To do this, we need to identify elements (and relations) of the target system in a way that does not presuppose the very model we are testing, and neuro-imaging is an important means for such identification. If neuro-imaging is to play this role, then we must assume that imaging can reveal causal structure;

otherwise, we could not use these images to assess the "fit" between model and target system. In the context of confirmation, we do not assume that the causal structure neuro-imaging reveals corresponds to the causal structure specified by the model. Suppose *pain* is an element in my model, and pain is defined as a state that satisfies a particular causal profile. If we use images to test this model, then there has to be a way of specifying the content of these images that makes no reference to pain. After all, if the image reveals that our model is inaccurate, then though the image reveals something—and thus has content—it won't have pain as its content (for otherwise it would reveal the model to be accurate).

However, suppose we are convinced that a model specified by a particular functional analysis is accurate. In that case, the model can provide an interpretative framework for "reading" images, e.g., we can say that an image depicts pain. How does this bear on the legal contexts in which these images might be used? It should be fairly clear that such contexts would *assume* the existence of well-confirmed models, models that guide our "reading" of images. If this is right, then one challenge to using such images in the court of law is the challenge of *transparency*: neuro-imaging can provide reliable information about the mind *provided* that we have a fairly detailed and well-confirmed account of the causal or functional roles that brain states play.

A second challenge to the use of neuro-imaging is what I will call the challenge of *uniqueness*. Suppose that I am a cognitive scientist who has provided incredibly detailed and well-confirmed functional analyses of a great number of a particular person's capacities, e.g., Nickey's capacities. As a scientist, I am not interested in providing functional analyses of only Nickey's capacities; I want my explanations to generalize. In the context of law, if our aim is to bring neuroscientific models to bear on the issue of whether a defendant possesses certain capacities, and if we can reasonably assume that Nickey is representative of the human population, then I might be justified inferring functional role from an image of the defendant's brain. But the assumption of representativeness is doubly problematic. First, suppose that executing a partial products algorithm explains Nickey's capacity to multiply (and this is well confirmed). Supposing that Noah also has the capacity to multiply, it would not follow that executing a partial products algorithm explains Noah's capacity. Noah might execute a successive addition algorithm, for example. If that were the case, then though Noah's brain would differ in important ways from Nickey's brain, those differences wouldn't entail a difference with respect to the capacity to multiply. Second, given the evidence for neural plasticity, Noah's brain might differ in important ways from Nickey's brain even if executing a partial products algorithm explains Noah's capacity to multiply and even if we are concerned with only those parts of the brain that implement the partial products algorithm. According to cognitive scientist Joan Stiles,

> Data from both animal and human studies show that plasticity plays a central role in the normal development of neural systems allowing for adaptation and response to both exogenous and endogenous input. The capacity for reorganization and change is a critical feature of neural development, particularly in the postnatal period. Subtractive processes play a major role in the shaping and sculpting of neural organization. However, plasticity is neither transient nor unique to developing organisms. With development, neural systems stabilize and optimal patterns of functioning are achieved. Stabilization reduces, but does not eliminate, the capacity of the system to adapt. As the system stabilizes, plasticity becomes a less prominent feature of neural functioning, but it is not absent from the adult system.[15]

For present purposes, the potential importance of plasticity is this: given the capacity for reorganization of which Stiles speaks, brains that are significantly different organizationally may nonetheless be functionally quite similar, i.e., the same functional analysis may be implemented in a variety of ways. The extent of this variation is of course an empirical matter, and I want neither to overstate nor understate what the evidence shows about plasticity. My goal here is merely to draw attention to the kinds of questions that need to be answered when we are considering the evidential value of neuro-imaging.

BEYOND CORRELATION

How we conceive the relevance of neuroscientific evidence to questions about mind will depend on which version of physicalism we adopt, and in this chapter I've considered the relevance of neuro-imaging through the lens of functionalism. However, it is worth comparing our results with the results one gets when using a different lens, e.g., the Rylean lens adopted by Pardo and Patterson. Pardo and Patterson have much to say about neuro-imaging, but—perhaps unsurprisingly—their perspective differs substantively from the one offered here. Let's start with the goal of finding neural *correlates* to mental states and capacities. Pardo and Patterson have no objection to this, and they concede that neural activity can provide inductive evidence that a person is in a particular mental state or possesses a particular capacity.[16] Functionalists can agree with Pardo and Patterson on this point, even if the correlations provide something weaker than identity (in the case of mental states) or explanation (in the case of capacities). As we've seen, neural activity is correlated with oxygenation levels, and oxygenation is a property of physical states that may (or may not) implement functions. If oxygenation levels correlate with capacities, then knowledge of oxygenation levels can provide evidence of capacities. However, oxygenation levels do not *explain* capacities, even if the oxygenation levels we detect reflect activity relevant to the explanation of a capacity. To explain the capacity, we need to know what is contributed by the activity. Oxygenation levels reveal levels of activity, not what is contributed by that activity. Note, too, that oxygenation levels can be relevant even if the activity they indicate contributes nothing to the explanation of a capacity. For example, suppose activity in brain region A is relevant to the explanation of capacity C. Further suppose that activity in brain region A causes activity in brain region B, but this effect is irrelevant to the explanation of capacity C (e.g., it is merely noise). If activity in region A is correlated with capacity C and activity in region B is correlated with activity in region A, then oxygenation levels that indicate activity in region B can indicate the presence of capacity C even though the activity in region B does not contribute anything relevant to the explanation of capacity C.

Functionalists think we can discover more than mere correlations, however, and it is to this which Pardo and Patterson object. Consider first the attempt to identify mental states with brain states. As we've seen, Pardo and Patterson think that a lot of mental talk is talk of capacities, e.g., they think that to ascribe knowledge to a person is to ascribe a capacity. Because capacities are not states, capacities cannot be brain states. However, Pardo and Patterson do not deny the existence of mental states, and so it might be thought that there is nothing objectionable about identifying mental states with brain states. While Pardo and Patterson cannot appeal to the capacity/state distinction to rule out the identification of mental states with brain states, they think the identification can be ruled out on other grounds:

> For example, suppose someone said to us, "I intend to rob the bank tomorrow, and robbing the bank is impossible" … . It is nonsensical, if taken literally, to both intend to perform an action and to believe the action is impossible. … Suppose, however, that having an intention *just was* having a particular brain state … it is perfectly coherent … to say "my brain is in BS, and robbing the bank is impossible." Because the combination of brain state and the impossibility of the task makes sense, but the intention and the impossibility do not, the BS and the intention are not identical.[17]

This argument clearly conflates concepts and properties. Consider the putative identity of pain with c-fiber stimulation. According to the analytic functionalist, this identity claim is a claim about properties, not concepts. The *concept* of c-fiber stimulation is not the concept of an inner state apt for producing writhing and screaming; if it were, then 'pain = c-fiber stimulation' would be analytic and a priori, and it's not. Whether a statement does or does not "make sense" depends on the concepts used, and thus while it is nonsensical to say "I am in pain, but I am not in a state that is apt to cause screaming and writhing," it is not nonsensical to say "My c-fibers are stimulated, but I am not in a state that is apt to cause screaming and writhing." This doesn't show that pain isn't c-fiber stimulation; it shows that the *concept* of pain is not the *concept* of c-fiber stimulation. But as Frege taught us, difference in concept does not entail difference in referent, so from the fact that the concept of pain is different from the concept of c-fiber stimulation, it does not follow that these concepts pick out different properties.[18]

Though difference in concept does not *entail* difference in referent, difference in concept is obviously compatible with difference in referent. At the heart of the dispute between physicalists and dualists is the issue of whether our concepts of the mental refer to objects or properties picked out by our concepts of the physical. Pardo and Patterson deny that mental states are brain states, and while this does not by itself imply dualism, the only physicalist-friendly alternative that can accommodate this is not very attractive. To see this, consider Semir Zeki and Oliver Goodenough's claim that "subjective mental states—the feeling of love, the appreciation of beauty, the desire to cheat, the effort to read the mind of others and much else besides—have particular neural correlates that can be studied with such precision as modern-day technology affords, and which future technology will improve upon."[19] Because correlation is not identity, dualists like David Chalmers can happily accept that there are neural correlates to subjective mental states.[20] Physicalists thus need something stronger than mere correlation, and this is what the identity theory provides, e.g., mental states *are* brain states. In identifying mental states with behavioral dispositions, the philosophical behaviorist can also provide something stronger. To avoid dualism while rejecting the identification of mental states with brain states, Pardo and Patterson thus seem committed to saying that mental states, including intentions, are behavioral dispositions. But if that is the case, then the argument they give against identifying mental states with brain states seems unmotivated; why not simply argue that mental states are really dispositions, not states, and thus not brain states? The answer, I suspect, is they recognize that this strong form of philosophical behaviorism is deeply implausible. Because it never occurs to them that they can preserve the "conceptual" link between mental states and behavior while at the same time identifying mental states with brain states, i.e., analytic functionalism is never presented as a live option, they are left with the seemingly impossible task of defending a view that is neither dualist, nor functionalist, nor behaviorist.

In the case of capacities, Pardo and Patterson are surely right that identities are not forthcoming. But their discussion of this point is skewed by their omission of an *explanatory* link between brain states and capacities. Pardo and Patterson concede that having a certain kind of brain may be necessary for having certain kinds of capacities, but apparently they don't think brains explain capacities in any stronger sense. This is correct, in a way: absent functional characterization, knowledge of the brain leaves us on the wrong side of Leibniz' Gap. But when we provide functional characterization, i.e., when we explain capacities via functional analysis, this allows us to make sense of the idea that functionally characterized states, components, or processes are implemented in brains, and if they are, it allows us to make sense of identifying such functionally characterized states, components, or processes with the neural structures that implement them. It's not the capacity we are trying to explain that gets identified with brain states, but rather the functionally characterized states, components, or processes whose organization explains the capacity. Because neuro-imaging can in principle reveal the brain states that play the relevant functional roles, neuro-imaging can reveal more than correlations; it can reveal identities.

An example will help to make the contrasting perspectives on neuro-imaging clearer. A number of neuro-imaging techniques are purportedly able to show whether a person possesses information about some event, e.g., whether a witness did or did not see the commission of a crime. One such technique, called "brain fingerprinting," uses electroencephalography (EEG) to measure a person's brain waves.[21] This technique is grounded in the claim that a brain wave called P300 is distinctive to recognition, so if this brain wave occurs when a person is presented with a stimulus, this shows that the person encountered the stimulus previously. To illustrate, suppose I want to know whether you've seen a particular painting before. If I show you the painting and detect the presence of P300 in your brain, then I can conclude that you have seen the painting before—even if you sincerely deny having seen it before. In light of this, we might be tempted to believe that knowledge of the painting is stored in your brain and neuro-imaging can detect such knowledge. According to Pardo and Patterson, however, such a belief would rest on "two problematic presuppositions. The first presupposition is that whether one remembers something may be identified with particular neurological states of the brain. The second presupposition is that someone's *retaining* an ability (e.g., remembering something) implies the *storage* of that ability in a particular location."[22] As you may be able to guess, the alleged problem with the first presupposition is that "memory is an ability, power, or capacity,"[23] and since capacities are not states, memory cannot be identified with a brain state. We encountered this argument in Chapter 6, and the response I gave there applies here. We can grant that memory is a capacity; our interest is in explaining it. How is it that people are able to display or report correctly what they have previously experienced? Here's an answer: there is a part of the brain that encodes information about the experience and this information can be accessed by other parts of the brain, including the parts of the brain that are responsible for displaying or reporting this information. Disruptions to the brain might make that information unavailable for display or report, and in that case, we might say the person lacks whatever capacity it is that constitutes memory. But this would not entail that the information is not there, and if it is there, neuro-imaging techniques might reveal it. Turning to the second problematic presupposition, note, first, that encoding information isn't a matter of storing an ability. For example, an accurate street map of Chicago "stores" lots of information about Chicago, but whatever this information is, it isn't an ability. This confusion is reflected in Pardo and Patterson's objection to the idea that memory is a matter of the brain storing and accessing information. They think this idea is implausible because it presupposes memory, e.g., people would have to remember how to access and interpret information. Moreover, because people don't typically see their brains, such information, if there, would not be available for "viewing."[24] However, while it may very well be that the above explanation is faulty, the conception of storing and accessing information that is at issue here does not presuppose what Pardo and Patterson find objectionable. The explanation of memory I'm proposing does not appeal to further activities performed by the *person*; rather, it is what Daniel Dennett calls a *sub-personal* explanation that appeals to the activities of the person's "parts."[25] As Dennett notes, some of the capacities we attribute to a person can be analyzed in terms of further capacities of the person, e.g., discriminating good apples from bad may be analyzed into checking for color and crispness.[26] At a certain point, however, the capacities of a person cannot be analyzed in terms of further personal-level capacities. Consider a person's capacity to distinguish pains from other sensations. The explanation of this capacity won't be given in terms of something else the person does, i.e., a person does not do anything *else* in order to distinguish pains.[27] It does not follow from this that the capacity to distinguish pains cannot be analyzed, but it does follow that any such analysis can be given only in terms of sub-personal activities. The need to distinguish personal and sub-personal levels of explanation is perhaps nowhere more clear than in the case of intelligence. For example, it would obviously be a nonstarter to explain intelligent capacities of people in terms of the intelligent capacities of an inner man (a homunculus); any acceptable analysis of the intelligent capacities of people must ultimately bottom out in capacities whose exercise requires no intelligence. Fodor provides a nice description of what a successful explanation might look like, at least in general outline:

> Intelligent behavior typically exploits a "cognitive architecture" constituted of *hierarchies* of symbol processors. At the top of such a hierarchy might be a quite complex capacity: solving a problem, making a plan, uttering a sentence. At the bottom, however, are only the sort of unintelligent operations that Turing machines can perform: deleting symbols, storing symbols, copying symbols,

> and the rest. Filling in the middle levels is tantamount to reducing—analyzing—an intelligent capacity into a complex of dumb ones; hence to a kind of explanation of the former.[28]

To illustrate what Fodor has in mind, imagine the following system. The system includes a camera that periodically takes pictures of its surrounding environment. Each picture taken is encoded as a digital file, e.g., a string of position-intensity pairs, and each file is stored in a box. The system includes numerous boxes, and each box is stamped with a date, e.g., 10/12/2013. The system also includes an updating calendar, and whenever a picture is taken, the current date is stamped on the digital file. One part of the system compares the date on the file to dates on boxes, and it places the digital file in the box whose date matches the date on the file. The system receives other input as well, including commands posed in English. The commands are typed on a keyboard. One command is for memory retrieval, and when this command is followed by the specification of a date, (e.g., MEMORY 10/12/2013), there is a part of the system that responds by scanning boxes until it finds a match. Another part of the system then opens the box, pulls out all the files, and sends those files to a printer. The printer scans the string of position-intensity pairs. Each pair consists of numerals. The first numeral causes the printer to move to a certain location on the paper, and the second numeral causes the printer to apply a certain shade of grey. The printer stops once it has scanned and responded to each pair in the string, i.e., once it has finished printing the picture.

There are four important things to know about this system. First, if memory is the capacity to display or report correctly what was previously experienced, then the system has memory. Second, though the explanation of this capacity cites accessing files from boxes, accessing files from boxes is not an activity performed by the system as a whole. Rather, the activity is performed by a part of the system. Moreover, this activity does not presuppose memory, i.e., neither the system as a whole nor any part of the system has to remember how to access files in order to access files. The procedure for accessing files from boxes is not itself something encoded in a file and accessed; that procedure is part of the system's fixed functional architecture. Third, the process that eventuates in the printed picture in no way supposes "interpreting" anything, if interpretation implies that parts of the system have to assign meaning or appreciate the meaning of something. The activity of the parts can be explained solely in terms of scanning and matching symbols based on the shape, or syntax, of those symbols. This doesn't entail that the files lack content or do not encode information; rather, the point is that those parts need not know what the contents are in order for them to store and retrieve such files. Fourth, though digital files encode the content of the original picture, accessing this content does not require that parts of the system "look" at pictures. In fact, the picture is never available for viewing by the parts of the system that store and retrieve files from boxes. If the question is "What does the *system* do that explains the system's ability to display or report correctly what it previously experienced?" our answer might be "Nothing." The explanation we gave did not appeal to further activities of the system; it appealed to the activity of the system's *parts*.

Pardo and Patterson write that "Ascriptions of 'deciding' and 'choosing' to the brain are examples of the mereological fallacy. The brain is not an agent inside the person making decisions and choices—decisions and choices that then cause the person to make internal, conscious decisions and choices about what to do."[29] In contexts where neural activity is part of a sub-personal explanation of personal-level capacities, it will indeed be a mistake to characterize that activity in personal-level terms, e.g., in terms of deciding and choosing. Do cognitive scientists make this mistake when characterizing the brain as an information processor? As Dennett notes, it is an easy one to make:

> In his purest form the little man in the brain takes on the guise of brain-writing reader, an intelligent, communicating system capable of understanding messages. Positing the brain-writing reader is almost irresistible, for if we cannot understand central states and events of the nervous system as bearing content, as being messages of some sort, it is not clear how we can understand them at all. The temptation must be resisted, however, by recognizing the disanalogies between verbal communication and non-verbal intra-cerebral communication and indeed the primacy of non-verbal communication. Other roles played by the little man in the brain are merely specialized roles projected inwards from the details of our initial analysanda, the variety of affairs of a person. The solitary audience in the theatre of consciousness, the internal decision-maker and source of volitions or directives, the reasoner, if taken as parts of a person, serve only to postpone analysis. The banishment of these concepts from our analysis forces the banishment as well of a variety of other self-defeating props, such as the brain-writing to be read, the mental images to be seen, the volitions to be ordered, and the facts to be known. These props are self-defeating because they could only serve the functions for which they were designed in conjunction with interior person-analogues, and hence as elements in an analysis they reproduce the problems like images in a hall of mirrors.[30]

But while invoking the little man would be a mistake, Dennett also argues that we do not need to invoke little men to make sense of the brain as information processor:

> People arrive at conclusions, and, as the bland verb suggests, this is not a process that people go through or an activity in which they engage, so we cannot

> ask the question "How do you arrive at a conclusion?" and expect an answer in the form "First I do this, and then I do that"; people do not do anything in order to arrive at conclusions, but their brains must … . Were we to take what goes on in the brain and analyse it into parts, we should not expect those parts to be, say, concluding or deducing operations, for that is to confuse levels, and yet some operations of a different sort must occur. When computers are made to perform logical operations, the abstract, timeless transformations, and operations of logic are realized in physical, temporal operations, and the production of results or conclusions takes time and energy.[31]

Dennett wrote these passages over 40 years ago, and since that time cognitive scientists have developed fairly sophisticated information processing models of a variety of phenomena, including perception and action.[32] A great number of these models are sub-personal accounts of the target phenomena, and many modelers are philosophically savvy enough to avoid the kind of mistakes highlighted by Pardo and Patterson. Thus while cognitive scientists are not immune from confusing personal and sub-personal levels of explanation, it is simply incorrect to claim that all information processing characterizations of neural activity commit the mereological fallacy.

The point of the toy example of memory is not to show how an actual person's memory works. Nor am I claiming that if information is stored in the brain, then our current imaging techniques reveal this information (to reveal this information, we need to solve the problem of transparency. Arguably, we are a long way off). I am merely addressing the "conceptual" objections that Pardo and Patterson raise for this example of neuro-imaging. More importantly, the example highlights the way that judgments about the potential significance of neuro-imaging depend on the version of physicalism one adopts. Of course, getting clear about the potential significance of neuro-imaging is not the same thing as showing its actual significance to the current practice of law, and I want to close this chapter by considering current practice through the perspective on neuro-imaging defended here.

ON A NEED TO KNOW BASIS

The law uses a variety of tests to determine whether scientific testimony or evidence is admissible, including the *Frye* test[33] and the *Daubert* test.[34] Common to these tests is the issue of whether the testimony or evidence meets standards set by the scientific community. In short, we want to know whether the scientific community would accept that the testimony or evidence shows what it purports to show. But questions surrounding the reliability of the inferences we draw from evidence matter only if the inferences are potentially relevant to the case at hand, and so the issue of relevance is as much a legal concern as it is a scientific one. Recall the case of Charles Whitman, the man who went on a murderous rampage at the University of Texas. Suppose Whitman hadn't died but was instead arrested and charged with multiple counts of first-degree murder, and further suppose we were able to probe Whitman's brain using current neuro-imaging techniques. Based on the results of imaging, it is conceivable that the scientific community would agree that Whitman was unable to control his behavior. But while such a conclusion might be scientifically sound, it might also be irrelevant. For example, suppose Whitman pleads not guilty by reason of insanity and the test for insanity used focuses solely on the capacity to appreciate the criminality of conduct. Because being unable to control behavior does not entail lacking the capacity to appreciate the criminality of conduct, the evidence about lack of control might prove moot. However, suppose that the court uses a wider test, one that factors in the ability to control behavior. I said it is conceivable that the scientific community would agree that Whitman was unable to control his behavior. But would they in fact agree? To answer this, we need to consider the extent to which the challenges of transparency and uniqueness are met in this case.

I drove a Honda CRX for most of the 1990s. It was a great car—fun, reliable—but as I was driving home one day the engine abruptly stopped working. I towed the car to a local garage and was told by the mechanic that the timing belt had snapped, resulting in extensive deformation to several valves. I didn't know much about cars, but I knew that valves are necessary for a running engine and that slight deformations can disrupt their proper functioning. I was thus justified in concluding, as I did, that my car's engine wouldn't—*couldn't*—run again in that condition (a conclusion that the mechanic confirmed). My knowledge about the relationship between valves and running engines wasn't merely correlational; I knew something about the contribution valves make to an engine's operation, e.g., they allow air in and out of cylinders. Moreover, I knew that valves are a design feature common to modern automobile engines, including engines that otherwise differ in important ways (e.g., piston vs. rotary engines). This suggests two things. First, a system need not be fully transparent in order to justify conclusions about a system's capacities. Second, there are certain design features about which the assumption of uniqueness can be plausibly established. Neuroscientists know many more details about brains than I know about cars, but a brain is also vastly more complicated than an automobile engine. In terms of relative ignorance, current scientific knowledge of brains may be comparable to my knowledge of automobile engines. But just as I don't need to be an expert on engine design to know that valves are a ubiquitous design feature and that deformed valves prevent an engine from running properly, there can be cases where our lack of knowledge of the brain does not prevent us from drawing reasonable conclusions about a person's capacities.

Recall that an autopsy of Whitman's brain revealed that a large tumor had impacted his amygdala, a condition that current imaging technologies would likely detect. Assuming such detection, should this kind of evidence be admitted at trial? We can perhaps work our way to an answer to this question by asking a series of other, increasingly refined questions. First, do we know enough about the amygdala to know that its absence would undermine behavioral control, and are current imaging techniques reliable enough to establish whether the amygdala is present or absent in a particular person's brain? Second, do we know enough about the functional role the amygdala plays in controlling behavior to know the kinds of disruptions to the amygdala that would undermine behavioral control, and are current imaging techniques reliable enough to establish whether such disruptions have occurred in a particular person's brain? The answer to these questions appears to be yes.[35] In short, there are no serious scientific grounds for doubting that the amygdala plays a crucial role in controlling behavior, or for doubting that the amygdala plays this role in humans generally, or for doubting that current neuro-imaging techniques can reliably detect at least *some* of the disruptions to function that compromise behavioral control. This does not imply that neuro-imaging would reveal such disruptions to Whitman's amygdala, of course; the point is just that neuro-imaging could reveal such disruptions if they were there. As such, there appears to be a plausible basis for arguing that this kind of evidence should be admitted.

Or rather, there appears to be a plausible basis for arguing that such evidence should be admitted only if we can also reasonably argue that the results of neuro-imaging tell us something about Whitman's amygdala at the time the murders were committed.[36] Assume that neuro-imaging reveals a disruption that compromises behavioral control. Still, because the neuro-imaging took place after the crimes were committed, it is possible that the disruption detected was not present at the time of those crimes. Two factors are particularly relevant to whether we can reasonably infer that the disruptions detected were present when the crime took place: the amount of time that elapsed between the commission of the crime and the use of imaging, and the growth rate of tumors. The car analogy can again help clarify the issues in play here. The mechanic who inspected my car's engine was not present at the moment the engine stopped running, but we can imagine circumstances that would render this fact unimportant. For example, my car was towed to a local garage, and towing a car does not impact the engine of the car towed (unless something goes quite wrong, which didn't happen). My car was placed in a bay upon arrival, and it remained untouched until the mechanic started the inspection an hour or so later. Given this, it is implausible to think that the valve deformation took place *after* the engine stopped running. Our knowledge of tumors might similarly render certain time gaps irrelevant. For example, if the neuro-imaging took place mere hours after the crimes were committed, then on the assumption that tumors of the size observed in Whitman's brain cannot develop in such a time span, we would be entitled to infer that the tumor we observe was present at the time of the crime.

Perhaps it will be objected that Whitman's case is atypical, and thus while it might be conceded that the aforementioned points about scientific consensus apply to that case, this does not give us good grounds for thinking that such consensus is to be found in a majority of the cases where the use of neuro-imaging is at issue. This is a fair point, and my discussion of Whitman is not intended to establish any general conclusions about the extent of consensus. If there is a lesson to be learned here, it is that we should avoid blanket conclusions about the evidential value of neuro-imaging. Instead, conclusions should be drawn on an individual basis and should reflect both our assessment of the particular technologies and techniques used to generate images and the degree of transparency needed to justify the functional ascriptions that are pertinent to the case at hand. Finally, we should not forget that if our concern is merely to establish the presence or absence of certain capacities, then not only might evidence of correlation prove sufficient (correlation without identity, or correlation without functional ascription), the value of such evidence might turn on the consilience achieved when other forms of evidence are considered, including behavioral evidence.

Much of the enthusiasm over neuro-imaging is generated by the hope that observing the brain will tell us something important about the mind. The slogan that *you are your brain* gives voice to this hope, and it does so precisely because it implies that something stronger than correlation is possible here. The slogan invites confusion, however, and we shouldn't let hope trump clear thinking. I've argued that functionalism best reflects the practice of contemporary cognitive science and provides a framework for moving from correlations to identities. In this way, functionalism captures the spirit of *you are your brain* while avoiding the conceptual pitfalls lurking around this phrase.

NOTES

1. Greene and Cohen, "Neuroscience Changes Nothing and Everything," 1779.
2. Ryle, *The Concept of Mind*, 32.
3. In what follows I talk of having or lacking a capacity. This is misleading insofar as it suggests that capacities are all or nothing, but this oversimplification does not affect the issues in play here.
4. One might plausibly wonder whether treating someone for schizophrenia (and perhaps it depends on the manner(s) of treatment) is even properly describable as

rehabilitation. One might think that rehabilitation worthy of the name requires a thoroughly participant-perspective approach, and that what I'm describing is *treatment* or *curing*, rather than the rehabilitation of a *person*. "Repair" or "capacity-restoration" or "treatment" may strike some as falling under a completely different heading, in terms of the perspective taken toward the offender, when compared to "rehabilitation" (for a suggestion along these lines, see Nicole Vincent's "Restoring Responsibility: Promoting Justice, Therapy, and Reform Through Direct Brain Interventions," *Criminal Law and Philosophy* 8 (1) (2014): 21–42). This is a worthy issue to explore, and if my use of "rehabilitation" here sounds misleading, substitute "treatment" or "cure." Thanks to Ty Fagan for bringing this point to my attention.

5. In an endnote Greene and Cohen remark that "it is conceivable that rationality could someday be redefined in neurocognitive rather than behavioural terms, much as water has been redefined in terms of its chemical composition. Were that to happen, neuroscientific evidence could then be construed as more direct than behavioural evidence" (Greene and Cohen, "Neuroscience Changes Nothing and Everything," 1784). However, it is fairly clear that this "redefinition" would amount to no more than the identification of whichever of the neural mechanisms that explain the relevant capacities. As such, this redefinition would simply be an application of the functionalist's strategy for deriving identities.

6. For a nice discussion of this issue, see Michael Perlin, "'His Brain Has Been Mismanaged with Great Skill': How Will Jurors Respond to Neuroimaging Testimony in Insanity Defense Cases?" *Akron Law Review* 42 (3) (2009): 885–916.

7. To forestall misunderstanding, let me be clear that the discussion in this section is *not* meant to imply anything about our current level of neuroscientific knowledge, nor is it meant to suggest that those working in the mind sciences operate in ignorance of the points I make. My aim, rather, is to make *explicit* both what is required to make well-founded inferences from images and the complexities involved in meeting those requirements.

8. Ronald Giere, *Scientific Perspectivism* (Chicago: University of Chicago Press, 2006).

9. Giere, *Scientific Perspectivism*, 41.

10. Giere, *Scientific Perspectivism*, 50–51.

11. Giere, *Scientific Perspectivism*, 55.

12. Patricia Churchland, *Brain-Wise* (Cambridge, MA: MIT Press), 3.

13. And as the analysis of the Planning Effect demonstrates (Chapter 5), such states can be part of a functional analysis of some capacity.

14. I say "some" because even if all mental states are brain states, it does not follow that all brain states are mental states.

15. Joan Stiles, "Neural Plasticity and Cognitive Development," *Developmental Neuropsychology* 18 (2) (2000): 237.

16. Pardo and Patterson, *Minds, Brains, and Law*, 10.

17. Pardo and Patterson, *Minds, Brains, and Law*, 138.

18. Gottlob Frege, "On Sense and Reference," in *Translations from the Philosophical Writings of Gottlob Frege*, ed. and trans. Peter Geach and Max Black (Oxford: Blackwell, 1980), 56–78.

19. Semir Zeki and Oliver Goodenough, "Law and the Brain: Introduction," *Philosophical Transactions of the Royal Society B: Biological Sciences* 359 (2004): 1661.

20. David Chalmers, "How Can We Construct a Science of Consciousness?" in *The Cognitive Neurosciences III*, ed. Michael Gazzaniga (Cambridge, MA: MIT Press, 2004), 1111–1120.

21. Lawrence Farwell and Sharon Smith, "Using Brain MERMER Testing to Detect Knowledge despite Efforts to Conceal," *Journal of Forensic Sciences* 46 (2001): 135–143.

22. Pardo and Patterson, *Minds, Brains, and Law*, 102.

23. Pardo and Patterson, *Minds, Brains, and Law*, 102.

24. Pardo and Patterson, *Minds, Brains, and Law*, 104.

25. Daniel Dennett, *Content and Consciousness* (New York: Routledge, 1969).

26. Dennett, *Content and Consciousness*, 104.

27. Dennett, *Content and Consciousness*, 103.

28. Fodor, *Psychosemantics*, 23.

29. Pardo and Patterson, *Minds, Brains, and* Law, 129.

30. Dennett, *Content and Consciousness*, 214–215.

31. Dennett, *Content and Consciousness*, 167.

32. A nice discussion of some of these models can be found in Clark, *Mindware*.

33. *Frye v. United States*, 293 F. 1013, 1014 (D.C. Cir. 1923). In *Frye*, the Court determined that the relevant test for admissibility is whether the evidence offered has gained general acceptance within the particular field in which it belongs.

34. *Daubert v. Merrill Dow Pharmaceuticals*, 509 U.S. 579, 593 (1993). In *Daubert*, the Court provided a four factor test for admissibility: (1) falsifiability; (2) error rates of the methods used; (3) peer review and publication of the relevant findings; (4) general acceptance within the scientific community.

35. For a detailed discussion of the theory and evidence that supports this conclusion, see Richard Redding, "The Brain-Disordered Defendant: Neuroscience and Legal Insanity in the Twenty-First Century," *American University Law Review* 56 (1) (2006): 51–127.

36. Insofar as we are concerned only with the charged crime. The evidence can also be relevant to the issue of whether Whitman is competent to stand trial.

Chapter 9

Does the Legal System Have a Diminished Capacity?

As implied by its name, the primary function of our criminal justice system is to provide justice, and central to this charge is meting out punishment to those who commit crimes. The connection between justice and punishment rests largely on retributivist principles, and our commitment to these principles rests on our more general commitment to the manifest image. We can thus agree with Stephen Morse that "there is no just punishment without desert and no desert without responsibility."[1] We should not let this commitment go unexamined, however, and much of the interest in neurolaw is generated by the uneasy feeling that advances in neuroscience have made—and will continue to make—this commitment untenable. The primary task of this book has been to quell that feeling, and by implication, to show that punishing criminals does not conflict with the aim of providing justice. But while the criminal justice system can in this sense perform its function, there are a number of challenges that the law must address if it is going to perform this function *well*. I divide these challenges into two kinds: internal and external. The internal challenges arise from the way the assumptions about responsibility and rationality that inform our legal theory are reflected in current legal practice. The external challenges arise from the way the criminal justice system operates within larger social, economic, and political structures. I take up these challenges in turn.

Among the internal challenges, two in particular stand out. The first is to provide tests or standards for responsibility that preserve the crucial link between responsibility and rationality—a point we met with in Chapter 7. Cognitive impairments can undermine rationality—and thus responsibility—and even if lacking substantial capacity to appreciate the criminality of conduct is not the only cognitive impairment relevant to rationality (and thus responsibility), the widely used M'Naghten Rule at least recognizes the importance of cognitive functioning to responsibility. But rationality is not a province of the understanding only; volitional impairments can similarly undermine rationality. Moreover, because volitional impairment can occur in the absence of cognitive impairment, we cannot assume that the M'Naghten Rule reliably screens for volitional impairment. Exclusive use of the M'Naghten Rule thus practically guarantees that people with volitional impairments will be found fully criminally responsible for their actions, a consequence that the law has at least sometimes recognized is deeply problematic.

While we should embrace standards and procedures that acknowledge both cognitive and volitional aspects of rationality, the failure to screen for volitional impairment is not the only weakness of the M'Naghten Rule. A successful insanity defense results in the defendant being found not guilty, and thus while a defendant who is deemed insane might still face incarceration (e.g., in a state-run mental hospital), such a person will not be deemed responsible. But what of an unsuccessful insanity defense? Consider the case of Mark Becker.[2] Becker was charged with first-degree murder after he shot and killed his former high school football coach. Becker suffered from schizophrenia at the time of the killing (a fact that the prosecution readily conceded), and the defense argued that because Becker had acted under the delusional belief that his former coach was the devil, Becker was not guilty by reason of insanity. The argument did not convince the jury, and Becker was ultimately sentenced to life in prison. Regardless of what one might think of the jury's decision, notice that there seems to have been just two options available: either Becker is insane and thus not responsible, or Becker is sane and thus fully responsible. It becomes clear that these were the only two options when we recall the basic rationale behind the (now widely discarded) diminished capacity plea. The doctrine of diminished capacity—known alternatively as the doctrine of diminished *responsibility*—provides that evidence concerning a defendant's mental state can result in a reduced charge, e.g., first-degree murder to manslaughter. The implication is that the offense with which one is charged is supposed to reflect the degree to which one is responsible (if convicted), so to be found guilty of first-degree murder is to be found fully responsible for the murder. But there is a significant flaw in this way of thinking about responsibility. Consider a person who, like Becker, is charged with first-degree murder, but who, unlike Becker, is found not guilty by reason of insanity. If we assume that both the charge and the verdict were justified, then we can be justified in charging a person with first-degree murder even if it turns out that the person lacks substantial capacity to appreciate the criminality of conduct. Now consider a person who, though she can appreciate the criminality of conduct, suffers from other sorts of cognitive or volitional impairments, e.g., she suffers from the sorts of impairments that the Supreme Court said were characteristic of mentally retarded persons. If a person who lacks substantial capacity to appreciate the criminality of conduct can be justifiably charged with first-degree murder, then surely a person who can appreciate the criminality of conduct but suffers from other cognitive or volitional impairments can also be justifiably charged with first-degree murder. Under M'Naghten, such a person would not be found legally insane (since she can appreciate the criminality of her conduct) and thus would be found fully responsible. In light of the *Atkins* ruling, this seems unfair. Exclusive reliance on the M'Naghten Rule is thus doubly problematic. First, it encourages us to reduce the two dimensions of rationality (understanding and will) to one (understanding). Second, it encourages us to treat responsibility as all or nothing (one is either fully responsible or not at all responsible).

Perhaps it will be replied that current law has mechanisms for avoiding these problematic results. For example, factors mitigating responsibility can be relevant during the sentencing phase even if they are not relevant during the trial phase, e.g., a person who suffers from various cognitive or volitional impairments might receive a lighter sentence than a person who does not suffer from such impairments. Suppose a person is found guilty of first-degree murder but, as a result of mitigating factors, receives only a 30-year sentence. If the relevance of the mitigating factors had to do with the way they bore on degree of responsibility, and if degree of responsibility is also reflected in the type of charge brought against a person, then the 30-year sentence might reflect what would have been appropriate had the person been convicted of a lesser charge (e.g., second-degree murder). In effect, the degree of responsibility is not automatically settled by the guilty verdict. Setting aside possible concerns about equity (How uniform are the standards used to determine mitigation? How consistently are they applied?), using the sentencing phase to determine degree of responsibility may conflict with public perceptions of responsibility. Imagine a person learns that Nickey and Noah were each convicted of first-degree murder. A few weeks later the person learns that while Nickey was sentenced to life in prison, Noah received only a 20-year sentence. I suspect this would strike the person as unfair. In learning of the guilty verdicts, the person would conclude that Nickey and Noah are each fully responsible for their respective crimes; each was charged with first-degree murder, after all. If Noah received a lighter sentence because he was found to be less than fully responsible, then it might appear to the person that the sentencing phase *overturned* what was determined during the trial phase. By contrast, if Noah had been found guilty of a lesser charge (e.g., second-degree murder), I suspect that the person would not find the difference in sentence unfair. Regardless of what we think about this particular example, however, the issue of public perception is important. If it turns out that everyday intuitions about responsibility are largely a function of the nature of the charged offense, then using the sentencing phase to determine degree of responsibility will likely lead to counterintuitive results.

If the issue of mitigation should be introduced during the trial phase, then it might be thought that the verdict category *guilty-but-mentally-ill* (GBMI) is a mechanism for doing this. In Chapter 7 I discussed clauses (a) and (b) of the Illinois state law regarding insanity. Omitted from that discussion was clause (c):

> A person who, at the time of the commission of a criminal offense, was not insane but was suffering from a mental illness, is not relieved of criminal responsibility for his conduct and may be found guilty but mentally ill.[3]

Several states now include similar provisions, but what is the purpose of this verdict category? Legally speaking, the qualifier "but mentally ill" carries no official weight. For example, while a GBMI verdict might factor into a mitigation argument given at sentencing, there is no *requirement* that the recognition of mental illness count as a mitigating factor. In short, it is up to the sentencing judge to decide whether the mental illness is a mitigating condition. Moreover, defendants who are found GBMI are not entitled to benefits that are not also entitled to any other prisoner, including mental health care. As Jennifer Bard notes, finding a defendant GBMI is thus like finding a defendant "guilty but hypertensive."[4]

A second internal challenge arises from how we use imprisonment. In Chapter 7 I noted that there are at least four grounds for incarceration—desert, deterrence, incapacitation, and rehabilitation—but among those grounds, only desert and deterrence seem to justify using incarceration for punitive purposes. If the aim of imprisonment is to punish, then it is unfair to imprison a person if the only justification for doing so is incapacitation or rehabilitation. Incarceration in any form deprives a person's liberty of movement, of course, and while imprisonment can achieve the aim of incapacitation (and perhaps rehabilitation, though this seems dubious), it is plausible to suppose that imprisonment deprives a person of more than what is required to incapacitate him. In a great number of cases it might be equally plausible to suppose that a person deserves punishment, so it looks like we can defend the additional deprivations imposed by prison in those cases where both incapacitation and desert justify incarceration. But an important complication arises when incarceration is "over determined" in this way, as we noted in our previous discussion of Three Strikes laws. Such laws impose longer sentences on repeat criminals, and if the longer sentences are justified on grounds of incapacitation, then the amount of time such a criminal ends up spending in prison will be more than what considerations of desert can justify. To put the point another way, while incapacitation might justify longer *incarceration* in these cases, incapacitation cannot by itself justify longer *imprisonment*. Similarly, because rehabilitation cannot justify punitive forms of incarceration, we cannot ground extra imprisonment in considerations of rehabilitation.

But what about desert or deterrence? For example, we might argue that repeat criminality makes a person deserving of a longer sentence. Or we might argue that the longer sentence serves to deter others. If either of these arguments is successful, then we can defend the extra time in prison even if incapacitation cannot provide that defense. However, each argument encounters a difficulty. Let's start with the idea that repeat criminals deserve longer sentences. Imagine a person is found guilty of robbing a liquor store, and because the person has previous felony convictions on his record, he is given a sentence that is longer than it would have been had he not been previously convicted. That the person was in a position to rob the liquor store implies that he had served whatever prison time he received for those previous convictions, and in having served this time (assuming he spent time in prison), there is the presumption that the person has paid "his due."[5] In having paid his due, we no longer view the person as deserving punishment. To say that the person deserves a longer sentence because of the person's past transgressions, we need to assume that there is some residual desert that can justify punishment beyond what is justified by the robbery alone. But it appears that there is none. Now consider deterrence. We might reasonably doubt that punishment is ever effective at deterring others, but even if we set aside this general doubt, there is an air of paradox in defending punishment on the grounds that punishment will deter repeat criminals. Being a repeat criminal provides prima facie evidence that such persons are not deterred by punishment, so while it is in principle possible that imposing longer sentences on repeat criminals would deter otherwise repeat criminals, the defender of deterrence has the burden of showing that it is plausible. Maybe the argument would go something like this: those prone to repeat criminality are deterred by penalties *stiffer* than those we typically impose for committing a crime, but because the penalties we typically impose deter most people from committing crimes, imposing stiffer penalties across the board would result in gratuitous levels of punishment. As a compromise, we reserve stiffer penalties for those who demonstrate that they are not deterred by the typical penalties. Perhaps an argument along these lines can be made to work;[6] for present purposes, it is enough to motivate the need for making such an argument. Unless we can appeal to desert or deterrence to justify the longer periods of incarceration that are mandated by Three Strikes laws, it will be hard to defend the idea that the extra time should be spent in prison.[7]

While the challenges raised by our use of imprisonment are in some important ways unrelated to the challenges raised by our use of the M'Naghten Rule, there are also notable ways in which the two are related. The recent rise of mental health courts provides both a good illustration of their relation and a useful model for thinking about how to address these challenges. In 2000, President Bill Clinton signed into law a bill that provides funding for the development of courts whose purpose is "to divert non-violent offenders into separate courts for people with mental illness ... as a move to decriminalize mental illness"[8] and whose promise is to recognize "that a person can be impaired by mental illness without being totally disabled by it."[9] There are at least two ways to read the underlying aim of these courts, but only one of them will concern me here. On the reading I am *not* concerned with, the purpose and promise of mental health courts is to recognize that a person might not be responsible for her actions even if she would qualify as sane under the M'Naghten Rule. This reading is suggested by the fact that treatment is offered as an *alternative* to imprisonment, and if the offender avoids imprisonment, it is plausible to view the offender as not responsible for her offense. In turn, this encourages us to read "impaired ... without being totally disabled" as implying that certain impairments can absolve one of responsibility even if those impairments do not prevent one from appreciating the criminality of conduct. The problem with this reading is that, while mental health courts might be an improvement to our current system, such courts will reinforce the idea that responsibility is an all or nothing affair. On the more promising reading, the one that I am concerned with, recognizing that a person can be impaired without thereby being totally disabled amounts to recognizing that responsibility is a matter of degree. Because the M'Naghten Rule invites us to treat responsibility as an all or nothing affair, criminals suffering from mental illness who are ushered through the normal court system are likely to be judged more responsible than they in fact are. In this way, mental health courts appear poised to provide more accurate assessments of responsibility. Regarding incarceration, the purpose of mental health courts is not to avoid punishment, but to divert offenders away from the forms of punishment imposed by the current prison system. A person may deserve punishment but not deserve the punishment she receives. This is a straightforward implication of a legal system like ours which acknowledges the legitimacy of both retribution and the Eighth Amendment prohibition on cruel and unusual punishment. In *Atkins*, the Court ruled that executing mentally retarded persons violated the Eighth Amendment. I've argued that holding responsible and taking responsibility are symmetrical in the sense that a justification for either is a justification for both, and the same applies to punishment that is based on desert. If a person's mental illness limits the extent to which she can meaningfully take responsibility, then it would be unfair to punish the person in the same way we would punish someone who is not so limited. In emphasizing treatment, mental health courts also acknowledge the fact that rehabilitation or treatment can provide grounds for incarceration. Setting aside the issue of whether the punishment administered by prisons is unfair, if prisons lack the resources needed for effective rehabilitation or treatment, then we must decide whether it is more important to punish or rehabilitate/treat. The need for mental health courts reflects the fact that rehabilitation or treatment can sometimes take priority over punishment.

While it is likely that mental health courts do not spring to mind when people think about the criminal justice system, the underlying rationale for such courts is also at work in the much older and more familiar arena of juvenile justice. Cornell Law School's Legal Information Institute defines juvenile justice as "the area of criminal law applicable to persons not old enough to be held responsible for criminal acts"[10] and notes that "The main goal of the juvenile justice system is rehabilitation rather than punishment."[11] While we should resist the suggestion that juveniles are in *no* way responsible for their actions, it is widely accepted inside and outside the legal community that juveniles typically do not bear the same responsibility for their actions as normal adults do. When we reflect on our reasons for this belief, it should be clear that it is not the age of the offender per se that matters, but rather what we assume age to indicate about the cognitive, affective, or volitional capacities of offenders. Put simply, we don't think that the cognitive, affective, or volitional capacities of the typical child or adolescent are sufficiently developed to make them full-standing members of the moral community, e.g., children have difficulty controlling impulses in much the way that some psychopaths, addicts, and persons with mental retardation do. But if these facts about juveniles explain—and justify—why we respond to juvenile offenders differently than we do normal adults, parity of reasoning enjoins us to respond differently to adult offenders who have similarly underdeveloped or compromised cognitive, affective, or volitional capacities. In this way, the juvenile justice system may provide a useful model for thinking about how to respond to those adult offenders who have a diminished capacity. It may also be that thinking about diminished capacity in this way can help mental health courts achieve the kind of legitimacy that the juvenile justice system largely enjoys.

Although establishing mental health courts can be correctly viewed as a way of responding to some of the internal challenges facing our legal system, the felt need for such courts is also a response to an external challenge confronting law. The way our current legal system treats criminals suffering from mental illness does not seem altogether fair, but while this fact exposes weaknesses within the current system, such weaknesses would likely not be viewed as presenting a pressing practical problem if the number of criminals suffering from mental illness were significantly lower. But given our current legal system, it is no wonder that the closing of large, state-run mental institutions over the last several decades has made imprisonment our de facto response to mental illness.[12] Practically, if not theoretically, such closures compel us to adopt the participant perspective toward those whom we previously adopted an objective attitude. I've spent a good amount of time arguing that it is a mistake to think physicalism and determinism require us to abandon the participant perspective, but defending the integrity of the participant perspective is not the same as defending its adoption toward all. We need to avoid both the Scylla of *no one is responsible* and the Charybdis of *everyone is responsible*. Navigating between them successfully will require us to get our conceptual house in order, but the difficulties we face are not merely a product of bad thinking. For example, while the claim that determinism implies that no one is responsible may be a dubious "philosophical" claim, the claim that everyone is responsible is simply a product of reconciling our criminal justice system with a policy decision to defund public mental institutions, i.e., this policy decision is not primarily motivated by philosophical commitments concerning free will and responsibility but rather judgments about how we should allocate resources. Although meeting the internal challenges faced by law might ameliorate some of the consequences of this policy decision, as Bard notes, by itself this may amount to little more than rearranging decks chairs on a sinking ship.[13] We should not be asking the law to wade into this morass, much less clear it up. Nonetheless, our legal system cannot function well until we do clear it up.

Another external challenge facing the legal system concerns the perceived legitimacy of the demands made by our laws. For a demand to be legitimate, the person of whom the demand is made must be able to meet it. But this is not enough. Recall the example of the employer who demands flattery from her employees. An employee may be able to meet this demand, but there is something about the *content* of the demand that makes this demand illegitimate. If an employee does not see herself as having a good reason to flatter her employer, then it won't make sense for her to take responsibility for failures to meet this demand. A well-functioning legal system is one in which people take responsibility for their failures to meet the demands of law, and thus a well-functioning legal system rests on the assumption that people see themselves as having a good reason to obey the laws. If a person does not see herself as having such a reason, then the legal system may end up holding this person responsible for a crime despite the fact that it makes no sense for the person to take responsibility for the crime. The significance of this point is perhaps nowhere clearer than in the way it bears on public discourse about race and responsibility. For example, consider the following remarks from President Barack Obama:

> We know that too many young men in our community continue to make bad choices. And I have to say, growing up, I made quite a few myself. Sometimes

> I wrote off my own failings as just another example of the world trying to keep a black man down. I had a tendency sometimes to make excuses for me not doing the right thing. But … we've got no time for excuses. Not because the bitter legacy of slavery and segregation have vanished entirely; they have not. Not because racism and discrimination no longer exist; we know those are still out there. It's just that in today's hyperconnected, hypercompetitive world, with millions of young people from China and India and Brazil—many of whom started with a whole lot less than all of you did—all of them entering the global workforce alongside you, nobody is going to give you anything that you have not earned.[14]

At stake here is the way the legacy of slavery, segregation, and racism impacts the issue of responsibility. President Obama seems to be suggesting that, while we should not deny the impact of this legacy on the choices African-Americans make, this legacy does not completely absolve African-Americans of responsibility. Perhaps more importantly, President Obama is suggesting that the issue of whether African-Americans are responsible for their bad choices is largely irrelevant; given today's world, African-Americans should take responsibility for their failures because doing so will make them more competitive. But there is another way we need to understand the impact of slavery, segregation, and racism on the issue of responsibility, a way that bears on whether we should call certain choices *bad*.

Imagine a black male growing up in the most impoverished urban area of the United States. Suppose that at the age of 13 this male is given the opportunity to join a local gang. The boy knows that joining the gang will mean dropping out of school and pursuing a life of crime. Should he join the gang, or should he stay in school and continue to obey the law? Suppose his school is underfunded and overcrowded; his family is poor and he has to work afternoons and weekends to help make ends meet; his neighborhood is riddled with drug use and violent crime, crime that often results in the deaths of innocent members of the community. Given all of this, what can he reasonably conclude about the consequences of each course of action? Joining the gang may lead to prison or death, but it also presents the chance to earn a significant and steady income. By contrast, staying in school and obeying the law may not lead to future opportunities for advancement, e.g., attending college might not be a realistic possibility for him. Moreover, he knows that he will be vulnerable to crime even if he does not join the gang. Given all this, joining the gang can emerge as the *best choice*. If this sounds perverse, I suspect that it is because we are envisioning situations in which there really are better alternatives. Imagine a white male growing up in the most affluent suburban area of the United States. Suppose that at the age of 13 this boy is faced with the same choice that faced the black male. Should the boy join the gang, or should he stay in school and continue to obey the law? In this situation, it is not hard to understand why joining the gang would be a poor choice.

If the black male's situation reflects the legacy of slavery, segregation, and racism, then that legacy is relevant to responsibility because it has created situations in which it might make more sense for a person to break the law than to obey it. But if it makes more sense for a person to break the law than to obey it, then we cannot reasonably expect such a person to take responsibility for breaking the law. Moreover, if the person's best choice makes him particularly vulnerable to imprisonment, then we can hardly fault him if he responds to imprisonment by declaring it "another example of the world trying to keep a black man down." To think otherwise is to insult the person's dignity as a rational agent. Like the challenge posed by the closure of publically funded mental institutions, this challenge cannot be adequately addressed simply by addressing the internal challenges facing law.

Obviously more can and should be said about the challenges I've identified, both internal and external. My aim in mentioning them is to draw attention to those issues that are especially critical to whether our criminal justice system works well, and while it is beyond the scope of this book to provide a satisfactory resolution to these issues, I hope the framework I've developed can provide a useful way to think about them. There is one more challenge I want to mention, however, one that is a bit nearer to the core concerns of this book. Empirical and technological advances in neuroscience promise to reveal precise details about the neural mechanisms underlying thought and emotion, and by implication, the capacities crucial to free will and responsibility. While we may not be able to predict what those details will reveal, we should prepare now for the possibility that they will reveal free will and responsibility to be more limited than we currently assume them to be. While part of this might include the realization that the number of people who are not at all responsible for their behavior is larger than we initially imagined, I suspect the greater part would involve the realization that the free and responsible among us are not *as* free and responsible as previously thought. Moreover, if the raison d'être of the practices of holding and taking responsibility is the existence of D-responsible agents—and it is—then we should heed the optimist's point that there might be means for achieving D-responsible agents that are more effective than those practices. While this might not lead to the complete abandonment of the participant perspective, it might lead us to accept that the efficacy of these practices can be improved through various interventions, including pharmaceutical and surgical interventions. This prospect may seem dehumanizing to many, and if taking *you are your brain* seriously requires taking this prospect seriously, then many may seek a return to dualism and libertarianism. To those who would so seek, I offer the words of Simon Blackburn:

> Dualism is often supposed to make possible freedom, dignity, human experience itself. It underwrites the big words: the kinds of words that get on banners … people fear the alternative. Are we reducing people, in all their living colourful complexity, to drab monochrome machines, conditioned into being this way or that, or worse, passive vehicles for our selfish genes? Absolutely not. The problem here is that the alternatives are posed as if they exhaust the field: either a free spirit, blissfully floating apart from the natural order, or a determined machine like a bus, or even a tram. … It is not the philosophy of compatibilism that denigrates human nature, but this way of putting the alternatives. This way of putting the matter supposes that nature is so awful that it takes a magical moment, a divine spark struck from the ghost in the machine, to make it sing. It is either clockwork (Zombies) or Ghosts. But that is the view that denigrates nature, including human nature.[15]

Though we may never fully complete the philosophical quest for self-understanding, a fair reckoning of the possibilities makes the prospects for doing so much brighter. But the issues we confront here are not merely speculative; they are also practical. Confronting them well will thus require our legal system, and the larger society of which it is a part, to be adaptive in ways characteristic of free and responsible agents. Our current system fails us in many ways, and we know it. The question now is whether we are equipped to respond effectively to these failures. If not, then our legal system has a diminished capacity indeed.

NOTES

1. Stephen Morse, "Excusing the Crazy: The Insanity Defense Reconsidered," *Southern California Law Review* 58 (3) (1985): 783.
2. Carlin Miller, "Iowa Coach Verdict: Jury Finds Mark Becker Guilty of Murder," *CBS News*, March 3, 2010, accessed October 1, 2016, http://www.cbsnews.com/news/iowa-coach-verdict-jury-finds-mark-becker-guilty-of-murder/.
3. "Illinois Compiled Statutes," Illinois General Assembly.
4. Jennifer Bard, "Re-arranging Deck Chairs on the Titanic: Why the Incarceration of Individuals with Serious Mental Illness Violates Public Health, Ethical, and Constitutional Principles and Therefore Cannot Be Made Right by Piecemeal Changes to the Insanity Defense," *Houston Journal of Health Law & Policy* 5 (2005): 142.
5. Strawson, "Freedom and Resentment," 91.
6. For evidence that it doesn't work, see John Worrall, "The Effect of Three-Strikes Legislation on Serious Crime in California," *Journal of Criminal Justice* 32 (4) (2004): 283–96.
7. In discussing prison and imprisonment, I do not mean to suggest that prison must have an exclusively punitive aim, i.e., I have no objection to using "prison" in

a more expansive sense that would include rehabilitation and incapacitation among its aims. Understood in this more expansive sense, a well-functioning prison system would be one in which the distinction between punitive and nonpunitive aims is honored and each of the aims is achieved. I insist only on two points. First, if the aim of our current prison system is understood in this expansive sense, then our prison system is clearly broken. Second, if the aim of our current prison system is exclusively punitive, then while not broken, it has significant shortcomings.

8. Bard, "Re-arranging Deck Chairs," 144–45.

9. Bard, "Re-arranging Deck Chairs," 145.

10. "Juvenile Justice," Legal Information Institute, accessed August 4, 2017, https://www.law.cornell.edu/wex/juvenile_justice.

11. "Juvenile Justice."

12. Ana Swanson, "A Shocking Number of Mentally Ill Americans End up in Prison Instead of Treatment," *The Washington Post*, April 30, 2015, accessed September 12, 2016, https://www.washingtonpost.com/news/wonk/wp/2015/04/30/a-shocking-number-of-mentally-ill-americans-end-up-in-prisons-instead-of-psychiatric-hospitals/.

13. Bard, "Re-arranging Deck Chairs."

14. Barack Obama, commencement address at Morehouse College, May 19, 2013, accessed September 12, 2016, https://www.whitehouse.gov/the-press-office/2013/05/19/remarks-president-morehouse-college-commencement-ceremony.

15. Blackburn, *Think*, 118.

Appendix

Court Cases and Legal Doctrine

Court Cases:

Frye v. United States, 293 F. 1013, 1014 (D.C. Cir. 1923).
Durham v. United States, 214 F.2d 862, 874–75 (D.C. Cir. 1954).
United States v. Brawner, 471 F.2d 969 (D.C. Cir. 1972).
People v. White, 117 Cal. App. 3d 270 (1981).
United States v. Hinckley, 672 F.2d 115 (D.C. Cir. 1982).
Daubert v. Merrill Dow Pharmaceuticals, 509 U.S. 579, 593 (1993).
Atkins v. Virginia, 536 U.S. 304 (2002).

Legal Doctrine:

Model Penal Code 4.01, American Law Institute (1962).
21 Am.Jur.2d *Criminal Law* 38 (1998).
40 Am.Jur.2d *Homicide* 109, n. 2 (1999).

References

American Society of Addiction Medicine. "Definition of Addiction." Accessed September 30, 2016, http://www.asam.org/quality-practice/definition-of-addiction.

Antony, Louise. "Everybody Has Got It: A Defense of Non-Reductive Materialism," in *Contemporary Debates in Philosophy of Mind*, edited by Brian McLaughlin and Jonathan Cohen, 143–159. Boston: Blackwell, 2007.

Appiah, Kwame. *Cosmopolitanism*. New York: W.W. Norton & Co., 2006.

Ariely, Dan. *Predictably Irrational*. New York: Harper Collins, 2009.

Aristotle. *Nicomachean Ethics*, translated by Terence Irwin, in *Readings in Ancient Greek Philosophy*, edited by S. Marc Cohen, Patricia Curd, and C.D.C Reeve. Indianapolis: Hackett, 2005.

Armstrong, David. *A Materialist Theory of the Mind*. London: Routledge, 1968.

Axelrod, Robert and William Hamilton. "The Evolution of Cooperation." *Science* 211 (1981): 1390–1396.

Ayer, A. J. *Philosophical Essays*. New York: St. Martin's Press, 1954.

Baier, Kurt. *The Moral Point of View*. Ithaca, NY: Cornell University Press, 1958.

Bard, Jennifer. "Re-arranging Deck Chairs on the Titanic: Why the Incarceration of Individuals with Serious Mental Illness Violates Public Health, Ethical, and Constitutional Principles and Therefore Cannot Be Made Right by Piecemeal Changes to the Insanity Defense." *Houston Journal of Health Law & Policy* 5 (2005): 101–179.

Bechara, Antoine, et. al. "Insensitivity to Future Consequences Following Damage to Human Prefrontal Cortex." *Cognition* 50 (1994): 7–15.

Bennett, Maxwell and Peter Hacker. *Philosophical Foundations of Neuroscience*. Malden, MA: Blackwell, 2003.

Bentham, Jeremy. *An Introduction to the Principles of Moral and Education*. London: Methuen, 1982.

Blackburn, Simon. *Think: A Compelling Introduction to Philosophy*. New York: Oxford University Press, 1999.

Block, Ned. "The Mind as the Software of the Brain," in *An Invitation to Cognitive Science* Vol. 3, edited by Daniel Osherson, et. al., 377–426. Cambridge, MA: MIT Press, 1995.

Braddon-Mitchell, David and Frank Jackson. *Philosophy of Mind and Cognition*. Malden, MA: Blackwell, 2008.

Carnap, Rudolf. *Philosophical Foundations of Physics: An Introduction to the Philosophy of Science*. New York: Basic Books, 1966.

Cartwright, Nancy. *The Dappled World*. Cambridge: Cambridge University Press, 1999.

Chalmers, David. "Facing Up to the Problem of Consciousness." *Journal of Consciousness Studies* 2 (3) (1995): 200–219.

———. "How Can We Construct a Science of Consciousness?" in *The Cognitive Neurosciences III*, edited by Michael Gazzaniga, 1111–1120. Cambridge, MA: MIT Press, 2004.

Churchland, Patricia. "Feeling Reasons," in *On the Contrary*, 231–254. Cambridge, MA: MIT Press, 1998.

———. *Brain-Wise*. Cambridge, MA: MIT Press, 2002.

Churchland, Patricia and Terrence Sejnowski. "Neural Representation and Neural Computation." *Philosophical Perspectives* 4 (1990): 343–382.

Churchland, Paul. "Eliminative Materialism and the Propositional Attitudes." *Journal of Philosophy* 78 (2) (1981): 67–90.

———. "Conceptual Similarity across Sensory and Neural Diversity: The Lepore/Fodor Challenge Answered." *Journal of Philosophy* 95 (1) (1998): 5–32.

———. *Plato's Camera*. Cambridge, MA: MIT Press, 2012.

Clapin, Hugh, et. al., ed. *Representation in Mind* Vol. 1*: New Approaches to Mental Representation*. Oxford: Elsevier, 2004.

Clark, Andy. *Being There*. Cambridge, MA: MIT Press, 1997.

———. *Mindware*. New York: Oxford University Press, 2014.

Contessa, Gabriele. "Scientific Representation, Interpretation and Surrogative Reasoning." *Philosophy of Science* 74 (1) (2007): 48–68.

Craver, Carl. "Functions and Mechanisms: A Perspectivalist Account," in *Functions: Selection and Mechanisms*, edited by Philippe Huneman, 133–158. Dordrecht: Springer, 2013.

Cummins, Robert. "Functional Analysis." *Journal of Philosophy* 72 (1975): 741–760.

———. *The Nature of Psychological Explanation*. Cambridge, MA: MIT Press, 1983.

———. *Representations, Targets, and Attitudes*. Cambridge, MA: MIT Press, 1996.

———. "'How Does It Work?' versus 'What Are the Laws?'" in *Explanation and Cognition*, edited by Frank C. Keil and Robert A. Wilson, 117–144. Cambridge, MA: The MIT Press, 2000.

———. "Neo-teleology," in *Functions: New Essays in the Philosophy of Psychology and Biology*, edited by Andre Ariew, et. al., 164–174. New York: Oxford University Press, 2002.

Cummins, Robert, et. al. "Epistemological Strata and the Rules of Right Reason." *Synthese* 141 (3) (2004): 287–331.

Cummins, Robert and Martin Roth. "Traits Have Not Evolved to Function the Way They Do Because of a Past Advantage," in *Contemporary Debates in Philosophy of Biology*, edited by Francisco Ayala and Robert Arp, 72–85. Malden, MA: Blackwell, 2010.

Damasio, Antonio. *Descartes' Error*. New York: Grosset/Putnam, 1994.

Dancy, Jonathan. *Practical Reality*. Oxford: Clarendon Press, 2000.

Davidson, Donald. "Actions, Reasons, and Causes." *Journal of Philosophy* 60 (23) (1963): 685–700.

———. "Truth and Meaning." *Synthese* 17 (1967): 304–323.

Dawkins, Richard. *The Selfish Gene*. Oxford: Oxford University Press, 1989.

Dennett, Daniel. *Content and Consciousness*. New York: Routledge, 1969.

———. *Brainstorms*. Cambridge: MIT Press, 1978.

———. *Consciousness Explained*. Boston: Little, Brown and Co., 1991.

———. *Freedom Evolves*. New York: Viking Press, 2003.

———. "My Brain Made Me Do It." Max Weber Lecture Series, European University Institute. Florence, Italy, December 15, 2010. Accessed August 30, 2016, http://cadmus.eui.eu/bitstream/handle/1814/16895/MWP_LS_2011_01.pdf?sequence=1.

Descartes, René. *Meditations on First Philosophy*, translated by John Veitch (1901). Accessed July 15, 2016, http://www.wright.edu/~charles.taylor/descartes/meditation4.html.

Duff, Antony. "Legal Punishment," in *The Stanford Encyclopedia of Philosophy*. Accessed August 4, 2017, https://plato.stanford.edu/entries/legal-punishment/.

Eagleman, David. "The Brain on Trial." *Atlantic*, July/August 2011. Accessed May 3, 2015, http://www.theatlantic.com/magazine/archive/2011/07/the-brain-on-trial/308520/.

Farwell, Lawrence and Sharon Smith. "Using Brain MERMER Testing to Detect Knowledge despite Efforts to Conceal." *Journal of Forensic Sciences* 46 (2001): 135–143.

Finlay, Stephen. "Oughts and Ends." *Philosophical Studies* 143 (3) (2009): 315–340.

Fischer, John Martin and Mark Ravizza. *Responsibility and Control: A Theory of Moral Responsibility*. New York: Cambridge University Press, 1998.

Fodor, Jerry. "Special Sciences." *Synthese* 28 (1974): 77–115.

———. *The Language of Thought*. New York: Crowell, 1975.

———. *The Modularity of Mind*. Cambridge, MA: MIT Press, 1983.

———. *Psychosemantics*. Cambridge, MA: MIT Press, 1987.

———. "A Modal Argument for Narrow Content." *Journal of Philosophy* 88 (1991): 5–26.

———. "Review of John McDowell's *Mind and World*," in *In Critical Condition*, 3–8. Cambridge, MA: MIT Press, 1998.

———. "Special Sciences: Still Autonomous after All These Years," in *In Critical Condition*, 9–24. Cambridge, MA: MIT Press, 1998.

Frankfurt, Harry. "Alternate Possibilities and Moral Responsibility." *Journal of Philosophy* 66 (23) (1969): 829–839.

Frege, Gottlob. "On Sense and Reference," in *Translations from the Philosophical Writings of Gottlob Frege*, edited and translated by Peter Geach and Max Black, 56–78. Oxford: Blackwell, 1980.

Giere, Ronald. *Explaining Science*. Chicago: University of Chicago Press, 1988.

———. *Science without Laws*. Chicago: University of Chicago Press, 1999.

———. *Scientific Perspectivism*. Chicago: University of Chicago Press, 2006.

Godfrey-Smith, Peter. "Folk Psychology as a Model." *Philosopher's Imprint* 5 (2005): 1–16.

Goodenough, Oliver. "Responsibility and Punishment: Whose Mind? A Response." *Philosophical Transactions of the Royal Society B: Biological Sciences* 359 (2004): 1805–1809.

Greene, Joshua and Jonathan Cohen. "For The Law, Neuroscience Changes Nothing and Everything." *Philosophical Transactions of the Royal Society B: Biological Sciences* 359 (2004): 1775–1785.

Grice, H.P. "Meaning." *Philosophical Review* 66 (1957): 377–388.

Hacking, Ian. *The Social Construction of What?* Cambridge, MA: Harvard University Press, 2000.

Hart, H.L.A. *Punishment and Responsibility*. New York: Oxford University Press, 1968.

Haugeland, John. "Representational Genera," in *Philosophy and Connectionist Theory*, edited by William Ramsey, et. al., 61–89. Hillsdale, NJ: Lawrence Erlbaum Associates, 1991.

Hempel, Carl. *Philosophy of Natural Science*. New Jersey: Prentice-Hall, 1966.

Hempel, Carl and Paul Oppenheim. "Studies in the Logic of Explanation." *Philosophy of Science* 15 (1948): 135–175.

Hume, David. *A Treatise of Human Nature*, edited by P.H. Nidditch. Oxford: Clarendon Press, 1978.

Horgan, Terence and James Woodward. "Folk Psychology is Here to Stay." *Philosophical Review* 94 (2) (1985): 197–226.

Illinois General Assembly. "Illinois Compiled Statutes." Accessed September 5, 2016, http://www.ilga.gov/legislation/ilcs/fulltext.asp?DocName=072000050K6–2.

James, Doris and Lauren Glaze. "Mental Health Problems of Prison and Jail Inmates." *Bureau of Justice Statistics* NCJ 213600 (2006).

Jones, Arthur Samuel. *Thoreau's Incarceration, as Told by His Jailer*. Berkeley Heights, NJ: Oriole Press, 1962

Kane, Robert. "Free Will: Ancient Disputes, New Themes," in *Reason and Responsibility*, edited by Joel Feinberg and Russ Shafer-Landau, 499–512. Belmont, CA: Wadsworth, 2002.

Kant, Immanuel. *The Critique of Practical Reason*, in *Great Books of the Western World* 42. Chicago: University of Chicago, 1952.

———. *The Philosophy of Law: An Exposition of the Fundamental Principles of Jurisprudence as the Science of Right*. Union, NJ: Lawbook Exchange, 2002.

Kim, Jaegwon. "Multiple Realization and the Metaphysics of Reduction," in *Supervenience and Mind*, 309–335. New York: Cambridge University Press, 1993.

———. *Supervenience and Mind*. New York: Cambridge University Press, 1993.

Kitcher, Philip. "Explanatory Unification and the Causal Structure of the World," in *Scientific Explanation*, edited by Philip Kitcher and Wesley Salmon, 410–505. Minneapolis: University of Minnesota Press, 1989.

———. *Science, Truth, and Democracy*. New York: Oxford University Press, 2001.

Kripke, Saul. *Naming and Necessity*. Oxford: Blackwell, 1980.

Laudan, Larry. "A Confutation of Convergent Realism." *Philosophy of Science* 48 (1981): 19–48.

Legal Information Institute. "Juvenile Justice." Accessed August 4, 2017, https://www.law.cornell.edu/wex/juvenile_justice.

Leibniz, Gottfried. *The Monadology*, in *Philosophical Essays*, translated by Roger Ariew and Daniel Garber, 213–224. Indianapolis: Hackett, 1989.

Levin, Michael. "A Compatibilist Defense of Moral Responsibility," in *Introduction to Philosophy*, edited by Louis Pojman and James Fieser, 455–466. New York: Oxford University Press, 2008.

Lewis, David. "An Argument for the Identity Theory." *Journal of Philosophy* 63 (1966): 17–25.

———. "Mad Pain and Martian Pain," in *Readings in the Philosophy of Psychology*, edited by Ned Block, 216–222. Cambridge, MA: Harvard University Press, 1980.

Lilienfeld, Scott and Hal Arkowitz. "What 'Psychopath' Means." *Scientific American*, December 1, 2007. Accessed October 3, 2016, https://www.scientificamerican.com/article/what-psychopath-means/.

Lycan, William. *Consciousness*. Cambridge, MA: MIT Press, 1987.

———. "Homuncular Functionalism Meets PDP," in *Philosophy and Connectionist Theory*, edited by William Ramsey, et. al., 259–286. Hillsdale, NJ: Erlbaum, 1991.

Maibom, Heidi. "The Mindreader as Scientist." *Mind and Language* 18 (2003): 296–315.

Marr, David. *Vision*. San Francisco: W.H. Freeman, 1982.

McDowell, John. *Mind and World*. Cambridge, MA: Harvard University Press, 1994.

McKenna, Michael. *Conversation and Responsibility*. New York: Oxford University Press, 2012.

McLaughlin, Brian. "Is Role-Functionalism Committed to Epiphenomenalism?" *Journal of Consciousness Studies* 13 (1–2) (2006): 39–66.

Menzies, Peter. "Reasons and Causes Revisited," in *Naturalism and Normativity*, edited by Mario de Caro and David Macarthur, 400–482. New York: Columbia University Press, 2010, Kindle edition.

Miller, Carlin. "Iowa Coach Verdict: Jury Finds Mark Becker Guilty of Murder." *CBS News*, March 3, 2010. Accessed October 1, 2016, http://www.cbsnews.com/news/iowa-coach-verdict-jury-finds-mark-becker-guilty-of-murder/.

Miller, David. "Popper's Qualitative Theory of Verisimilitude." *British Journal for the Philosophy of Science* 25 (1974): 166–177.

Millikan, Ruth. *Varieties of Meaning*. Cambridge, MA: MIT Press, 2004.

Moran, Richard. *Knowing Right from Wrong: The Insanity Defense of Daniel McNaughtan*. New York: Free Press, 1981.

Morse, Stephen. "Excusing the Crazy: The Insanity Defense Reconsidered." *Southern California Law Review* 58 (3) (1985): 777–836.

———. "New Neuroscience, Old Problems," in *Neuroscience and the Law: Brain, Mind, and the Scales of Justice*, edited by Brent Garland, 157–198. New York: Dana Press, 2004.

Morse, Stephen and Adina Roskies, ed. *A Primer on Criminal Law and Neuroscience*. New York: Oxford Series in Neuroscience, Law, and Philosophy, 2013.

Murphy, Jeffrie. "Legal Moralism and Retribution Revisited." *Criminal Law and Philosophy* 1 (2007): 5–20.

Nadelhoffer, Thomas, ed. *The Future of Punishment*. New York: Oxford Series in Neuroscience, 2013.

Nagel, Ernest. *The Structure of Science*. New York: Harcourt Brace & Co, 1961.

National Hurricane Center. "Hurricanes in History." Accessed October 1, 2016, http://www.nhc.noaa.gov/outreach/history/#new.

New York Times. "A Challenge to New York's Broken Parole Board," June 13, 2016. Accessed July 22, 2016, http://www.nytimes.com/2016/06/13/opinion/a-challenge-to-new-yorks-broken-parole-board.html?_r=0.

Newell, Alan and Herbert Simon. "Computer Science as Empirical Inquiry: Symbols and Search." *Communications of the Association for Computing Machinery* 19 (1976): 113–126.

Newton-Smith, William. *The Rationality of Science*. London: Routledge & Kegan Paul, 1981.

Nietzsche, Friedrich. *On the Genealogy of Morals*, translated by Walter Kaufmann and R.J. Hollingdale. New York: Vintage Books, 1989.

Nisbett, Richard and Timothy Wilson. "Telling More Than We Can Know: Verbal Reports on Mental Processes." *Psychological Review* 84 (3) (1977): 231–259.

Obama, Barack. Commencement Address at Morehouse College, May 19, 2013. Accessed September 12, 2016, https://www.whitehouse.gov/the-press-office/2013/05/19/remarks-president-morehouse-college-commencement-ceremony.

O'Connor, Timothy. "Free Will." *Stanford Encyclopedia of Philosophy*. Accessed June 3, 2016, http://plato.stanford.edu/entries/freewill/#1.3.

Oppenheim, Paul and Hilary Putnam. "The Unity of Science as a Working Hypothesis," in *Minnesota Studies in the Philosophy of Science Vol. 2*, edited by Herbert Feigl, et. al., 3–36. Minneapolis: Minnesota University Press, 1958.

Pardo, Michael and Dennis Patterson. *Minds, Brains, and Law: The Conceptual Foundations of Law and Neuroscience*. New York: Oxford University Press, 2013. Kindle edition.

Paternoster, Raymond. "How Much Do We Really Know about Criminal Deterrence?" *Journal of Criminal Law and Criminology* 100 (3) (2010): 765–824.

Perlin, Michael. "On 'Sanism.'" *Southern Methodist University Law Review* 46 (1992): 373–407.

———. "'His Brain Has Been Mismanaged with Great Skill': How Will Jurors Respond to Neuroimaging Testimony in Insanity Defense Cases?" *Akron Law Review* 42 (3) (2009): 885–916.

Piccinini, Gualtiero and Carl Craver. "Integrating Psychology and Neuroscience: Functional Analyses as Mechanism Sketches." *Synthese* 183 (3) (2011): 283–311.

Pinker, Steven. *How the Mind Works*. New York: W.W. Norton & Company, 1997.

Plato. *Euthyphro*, translated by G.M.A Grube, in *Plato: Five Dialogues*, edited by John Cooper, 1–20. Indianapolis: Hackett, 2002.

Poirier, Pierre and Robert Cummins. "Representation and Indication," in *Representation in Mind* Vol. 1*: New Approaches to Mental Representation*, edited by Hugh Clapin, et. al., 21–40. Oxford: Elsevier, 2004.

Popper, Karl. *Objective Knowledge*. Oxford: Oxford University Press, 1972.

Poundstone, William. *Prisoner's Dilemma: John von Neumann, Game Theory, and the Puzzle of the Bomb*. New York: Doubleday Anchor, 1992.

Putnam, Hilary. "Philosophy and Our Mental Life," in *Mind, Language, and Reality* Vol. 2, 291–303. New York: Cambridge University Press, 1975.

Pylyshyn, Zenon. "Complexity and the Artificial Study of Human Intelligence," in *Philosophical Perspectives in Artificial Intelligence*. Atlantic Highlands, NJ: Humanities Press, 1979.

Railton, Peter. "The Affective Dog and Its Rational Tale: Intuition and Attunement." *Ethics* 124 (4) (2014): 813–859.

Raz, Joseph. *Practical Reasoning and Norms*. London: Hutchinson & Co., 1975.

Redding, Richard. "The Brain-Disordered Defendant: Neuroscience and Legal Insanity in the Twenty-First Century." *American University Law Review* 56 (1) (2006): 51–127.

Rosen, Jeffrey. "The Brain on the Stand." *New York Times Magazine*, March 11, 2007. Accessed June 2, 2015, http://www.nytimes.com/2007/03/11/magazine/11 Neurolaw.t.html.

Rosenberg, Alexander. *Philosophy of Social Science*. Boulder, CO: Westview Press, 2012.

Roth, Martin. "Representation, Philosophical Issues About." *Wiley Interdisciplinary Reviews: Cognitive Science* 1 (1) (2010): 32–39.

———. "Folk Psychology as Science." *Synthese* 190 (17) (2013): 3971–3982.

Roth, Martin and Robert Cummins. "Neuroscience, Psychology, Reduction, and Functional Analysis," in *Explanation and Integration in Mind and Brain Science*, edited by David Kaplan. New York: Oxford University Press (forthcoming).

———. "Two Tales of Functional Explanation." *Philosophical Psychology* 27 (6) (2014): 773–788.

Rupert, Robert. "Functionalism, Mental Causation, and the Problem of Metaphysically Necessary Effects." *Nous* 40 (2) (2006): 256–283.

Ryle, Gilbert. *The Concept of Mind*. London: Hutchinson, 1949.

Sapolsky, Robert. "The Frontal Cortex and the Criminal Justice System." *Philosophical Transactions of the Royal Society B: Biological Sciences* 359 (2004): 1787–1796.

Scanlon, T.M. *What We Owe to Each Other*. Cambridge, MA: Harvard University Press, 1998.

Sellars, Wilfrid. "Empiricism and the Philosophy of Mind," in *Science, Perception, and Reality*, 127–198. Atascadero, CA: Ridgeview Publishing Company, 1963.

———. "Philosophy and the Scientific Image of Man," in *Science, Perception, and Reality*, 1–40. Atascadero, CA: Ridgeview Publishing Company, 1963.

———. "Fatalism and Determinism," in *Freedom and Determinism*, edited by Keith Lehrer, 141–174. New York: Random House, 1966.

Sifferd, Katrina, et. al. "Legal Insanity and Executive Function," in *The Insanity Defense: Multidisciplinary Views on Its History, Trends, and Controversies*, edited by Mark White, 215–242. Santa Barbara, CA: Praeger, 2017.

Smith, Michael. *The Moral Problem*. Oxford: Blackwell, 1994.

Sober, Elliot. "The Multiple Realizability Argument against Reductionism." *Philosophy of Science* 66 (4) (1999): 542–564.

Stace, Walter. *Religion and the Modern Mind*. London: MacMillan & Co., 1953.

Stiles, Joan. "Neural Plasticity and Cognitive Development." *Developmental Neuropsychology* 18 (2) (2000): 237–272.

Strawson, Galen. "The Impossibility of Moral Responsibility." *Philosophical Studies* 75 (1–2) (1994): 5–24.

Strawson, P.F. "Freedom and Resentment," in *Free Will*, edited by Gary Watson, 72–93. Oxford: Oxford University Press, 2003.

Swanson, Ana. "A Shocking Number of Mentally Ill Americans End up in Prison Instead of Treatment." *The Washington Post*, April 30, 2015. Accessed September 12, 2016, https://www.washingtonpost.com/news/wonk/wp/2015/04/30/a-shocking-number-of-mentally-ill-americans-end-up-in-prisons-instead-of-psychiatric-hospitals/.

Swoyer, Chris. "Structural Representation and Surrogative Reasoning." *Synthese* 87 (1991): 449–508.

Taylor, Richard. "Libertarianism: Defense of Free Will," in *Introduction to Philosophy: Classical and Contemporary Readings*, edited by James Fieser and Louis Pojman, 401–407. New York: Oxford University Press, 2008.

Teller, Paul. "How We Dapple the World." *Philosophy of Science* 71 (2004): 425–447.

———. "Provisional Knowledge," in *Constituting Objectivity*, edited by Michel Bitbol, et. al., 503–514. Netherlands: Springer, 2009.

Trivers, Robert. "The Evolution of Reciprocal Altruism." *Quarterly Review of Biology* 46 (1971): 35–57.

van Fraassen, Bas. *Laws and Symmetry*. Oxford: Oxford University Press, 1989.

van Inwagen, Peter. "The Powers of Rational Beings: Freedom of the Will," in *Philosophy: The Quest for Truth*, edited by Louis Pojman and Lewis Vaughn, 415–428. New York: Oxford University Press, 2014.

Vargas, Manuel. *Building Better Beings*. New York: Oxford University Press, 2013.

Vincent, Nicole. "Responsibility: Distinguishing Virtue from Capacity." *Polish Journal of Philosophy* 3(1) (2009): 111–126.

——— ed. *Neuroscience and Legal Responsibility*. New York: Oxford Series in Neuroscience, 2013.

———. "Restoring Responsibility: Promoting Justice, Therapy, and Reform Through Direct Brain Interventions." *Criminal Law and Philosophy* 8(1) (2014): 21–42.

Wegner, Daniel. *The Illusion of Conscious Will*. Cambridge, MA: MIT Press, 2002.

Wittgenstein, Ludwig. *On Certainty*, translated by Denis Paul and G.E.M. Anscombe. New York: Harper & Row, 1969.

Worrall, John. "The Effect of Three-Strikes Legislation on Serious Crime in California." *Journal of Criminal Justice* 32 (4) (2004): 283–296.

Yaffe, Gideon. "Book Review: Responsibility and Control: A Theory of Moral Responsibility." *Erkenntnis* 53 (2000): 429–434.

Zeki, Semir and Oliver Goodenough. "Law and the Brain: Introduction." *Philosophical Transactions of the Royal Society B: Biological Sciences* 359 (2004): 1661–1665.

Zimring, Franklin, et. al. *Punishment and Democracy: Three Strikes and You're Out in California*. New York: Oxford University Press, 2001.

Index

About the Author

Martin Roth is an associate professor of philosophy at Drake University. He received his PhD in philosophy from the University of California, Davis, and his research is primarily in the philosophy of mind and philosophy of science, with a special interest in questions surrounding scientific explanation, mental representation, and the nature of knowledge. His work has appeared in several journals and anthologies, including *Philosophical Studies*, *Mind and Language*, *Synthese*, and *Philosophical Psychology*.